SpringerBriefs in Applied Sciences and Technology

Manufacturing and Surface Engineering

Series Editor

Joao Paulo Davim ⓘ, Department of Mechanical Engineering, University of Aveiro, Aveiro, Portugal

W0037862

This series fosters information exchange and discussion on all aspects of manufacturing and surface engineering for modern industry. This series focuses on manufacturing with emphasis in machining and forming technologies, including traditional machining (turning, milling, drilling, etc.), non-traditional machining (EDM, USM, LAM, etc.), abrasive machining, hard part machining, high speed machining, high efficiency machining, micromachining, internet-based machining, metal casting, joining, powder metallurgy, extrusion, forging, rolling, drawing, sheet metal forming, microforming, hydroforming, thermoforming, incremental forming, plastics/composites processing, ceramic processing, hybrid processes (thermal, plasma, chemical and electrical energy assisted methods), etc. The manufacturability of all materials will be considered, including metals, polymers, ceramics, composites, biomaterials, nanomaterials, etc. The series covers the full range of surface engineering aspects such as surface metrology, surface integrity, contact mechanics, friction and wear, lubrication and lubricants, coatings an surface treatments, multi-scale tribology including biomedical systems and manufacturing processes. Moreover, the series covers the computational methods and optimization techniques applied in manufacturing and surface engineering. Contributions to this book series are welcome on all subjects of manufacturing and surface engineering. Especially welcome are books that pioneer new research directions, raise new questions and new possibilities, or examine old problems from a new angle. To submit a proposal or request further information, please contact Dr. Mayra Castro, Publishing Editor Applied Sciences, via mayra.castro@springer.com or Professor J. Paulo Davim, Book Series Editor, via pdavim@ua.pt.

Tanveer Saleh · Mir Akmam Noor Rashid ·
Wan Ahmad Bin Wan Azhar · Wazed Ibne Noor

Laser-MicroEDM Based Hybrid Micromachining

Microdrilling and Micromilliling

 Springer

Tanveer Saleh
Autonomous Systems and Robotics
Research Unit (ASRRU)
Department of Mechatronics Engineering
International Islamic University Malaysia
(IIUM)
Kuala Lumpur, Malaysia

Mir Akmam Noor Rashid
Autonomous Systems and Robotics
Research Unit (ASRRU)
Department of Mechatronics Engineering
International Islamic University Malaysia
(IIUM)
Kuala Lumpur, Malaysia

Wan Ahmad Bin Wan Azhar
Autonomous Systems and Robotics
Research Unit (ASRRU)
Department of Mechatronics Engineering
International Islamic University Malaysia
(IIUM)
Kuala Lumpur, Malaysia

Wazed Ibne Noor
Department of Mechanical Engineering
University of Creative Technology
Chittagong
Chattogram, Bangladesh

Autonomous Systems and Robotics
Research Unit (ASRRU)
Department of Mechatronics Engineering
International Islamic University Malaysia
(IIUM)
Kuala Lumpur, Malaysia

ISSN 2191-530X   ISSN 2191-5318 (electronic)
SpringerBriefs in Applied Sciences and Technology
ISSN 2365-8223   ISSN 2365-8231 (electronic)
Manufacturing and Surface Engineering
ISBN 978-981-97-8373-1   ISBN 978-981-97-8374-8 (eBook)
https://doi.org/10.1007/978-981-97-8374-8

This Springer imprint is published by the registered company Springer Nature Singapore Pte Ltd.
The registered company address is: 152 Beach Road, #21-01/04 Gateway East, Singapore 189721, Singapore

If disposing of this product, please recycle the paper.

Contents

Chapter 1
Introduction

Micromachining, a material processing technology, employs a subtractive method to generate features within the range of 1–500 μm, as noted by Masuzawa [1]. This versatile technique finds widespread application in various industries, including biomedical, automotive, aerospace, semiconductor, MEMS, and consumer electronics, driven by the escalating demand for product miniaturization. Broadly speaking, micromachining methods can be classified into two categories: tool-based micromachining and beam-based micromachining.

Tool-based micromachining involves the utilization of a physical tool that interacts with the workpiece through various physical, electrical, thermal, or chemical means, resulting in the engraving of the tool's shape onto the workpiece. Examples of such micromachining techniques encompass micromilling, micro electrodischarge machining (μEDM), and micro electrochemical machining (μECM).

In contrast, beam-based micromachining employs beams, such as focused ion beam (FIB), electron beam (EBM), and laser beam (LBMM), to eliminate material from the workpiece primarily through thermal interactions between the beam and the workpiece.

1.1 Laser Beam Micromachining

Laser beam micromachining (LBMM) is a widely used beam-based micromachining techniques with some unique features such as fast machining process as well as applicability on wide varieties of materials. In the subsequent sections the discussion on LBMM's history, principle and applications will be continued.

© The Author(s), under exclusive license to Springer Nature Singapore Pte Ltd. 2025   1
T. Saleh et al., *Laser-MicroEDM Based Hybrid Micromachining*,
Manufacturing and Surface Engineering, https://doi.org/10.1007/978-981-97-8374-8_1

1.1.1 History of Laser Material Processing

The term "laser" is an abbreviation for "light amplification by stimulated emission of radiation". This groundbreaking concept was initially proposed by Albert Einstein in 1917, laying the theoretical foundation for this remarkable technology. However, it wasn't until several decades later that significant advancements were made in its practical realization. In 1954, Charles Hard Townes of Columbia University spearheaded the research efforts toward achieving the first successful demonstration of stimulated emission in the microwave region, leading to the development of the maser. Building upon this achievement, Nicolaas Bloembergen of Harvard University further advanced the field by working on solid-state masers, which offered new possibilities and applications. A major milestone in laser technology came in 1960 when Theodore H. Maiman successfully constructed the world's first laser using a synthetic ruby crystal. This remarkable achievement marked the birth of the laser as we know it today, opening up a vast array of applications in various fields. One of the most significant developments in laser technology occurred in 1963 when Kumar Patel developed the CO_2 laser. This type of laser quickly emerged as a dominant force within the industry, and for the next fifty years, it played a crucial role in numerous applications and industries. In the 1980s, a significant breakthrough came with the introduction of small and affordable lasers, such as the Carbon Dioxide Slab Laser. These lasers revolutionized laser material processing by expanding its capabilities beyond traditional metals to include organic materials like plastic and rubber. This breakthrough paved the way for new possibilities and applications, impacting industries such as manufacturing, medicine, communications, and more. The evolution of laser technology from its theoretical foundations to practical realizations has truly been a remarkable journey. Today, lasers continue to push the boundaries of what is possible, driving innovation and shaping the world around us in countless ways [2–4].

1.1.2 Principle of Laser Material Processing

Lasers are widely used for material processing, especially in the field of micromachining, due to their ability to offer a diverse range of wavelengths (ranging from ultraviolet to infrared), pulse durations (ranging from femtoseconds to microseconds), and repetition rates (ranging from single pulses to Megahertz). These versatile attributes enable high-resolution material processing and micromachining in terms of both depth and lateral dimensions. Laser micromachining can be achieved through two primary methods: direct laser writing (DLW) and masked projection method (MP) [5].

The DLW method, illustrated in Fig. 1.1 [5], operates by focusing and scanning the incident laser beam over the workpiece in the X and Y directions using two rotating mirrors controlled by motors (as depicted in Fig. 1.1). Prior to reaching the

Fig. 1.1 Concept of direct laser writing [5]

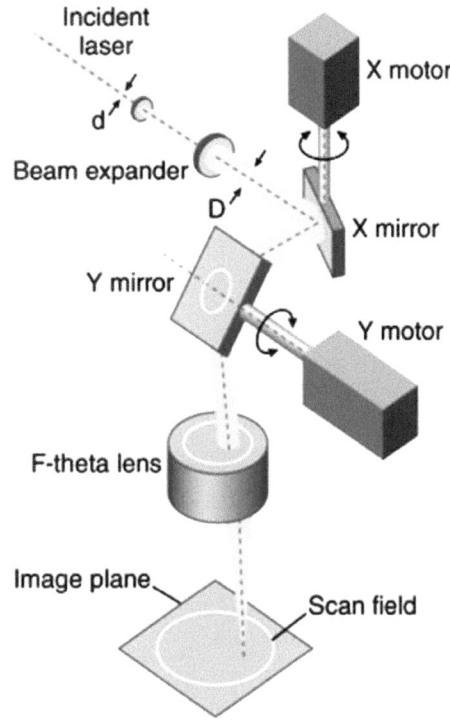

workpiece, the laser beam passes through a specialized laser scanning lens known as an F-Theta lens. An alternative configuration of the DLW system is presented in Fig. 1.2, wherein the laser source is directly mounted on the *X/Y* gantry, and the laser beam is scanned over the workpiece by the movement of *X* and *Y* stages.

Figure 1.3 [5] illustrates the working principle of the masked projection method (MP) employed in the LBMM process. In this method, a stationary laser beam is projected onto the workpiece through a pre-fabricated mask and appropriate optics, which ensure collimation and homogenization of the laser beam. One advantage of the DLW technique is that it eliminates the need for a pre-fabricated mask, which would otherwise require an additional expensive fabrication process. However, the dynamic behavior of the *X/Y* gantry scanning system or the *X/Y* mirror-based galvanometer scanning head may introduce laser beam instability, and the DLW process may also be time-consuming since the laser needs to be scanned over the entire workpiece. In contrast, the MP method of laser beam micromachining offers potential advantages such as enhanced stability and a potentially faster process due to the stationary laser beam and the use of pre-fabricated masks.

Overall, lasers for material processing, particularly in micromachining applications, provide a wide range of options in terms of wavelength, pulse duration, and repetition rates. The DLW and MP methods offer different approaches to laser beam micromachining, each with its own advantages and considerations. The choice

Fig. 1.2 XY gantry-based
DLW system (conceptual
illustration)

Fig. 1.3 Masked projection
method for the LBMM
process [5]

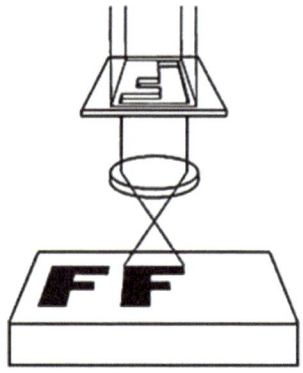

between these methods depends on the specific requirements of the application, such
as the desired resolution, processing speed, and cost constraints.

1.1.3   Application of Laser Beam Micromachining (LBMM)

LBMM employs both short and ultrashort laser pulses from gas or solid-state lasers
to selectively remove material. Ultrashort pulses are defined as those in which the
thermal diffusion depth is equal to or less than the optical penetration depth. The
LBMM process utilizes a wide range of lasers, spanning from ultraviolet (D-UV)

to mid-infrared (M-IR) wavelengths [6]. In practical applications of LBMM, solid-state lasers are more commonly used due to their ability to generate short pulses ranging from femtoseconds (fs) to nanoseconds (ns). Conversely, gas lasers can only achieve the shortest pulses in the nanosecond range [6]. Researchers and engineers are particularly interested in ultrashort laser sources for the LBMM process because they result in minimal heat affected zones (HAZ) and recast layers on the machined surface. However, the main limitation of using ultrashort laser sources (fs, ps) for LBMM is their high initial cost and limited penetration depth [7]. LBMM finds applications in various fields, including biomedical, photonics, microelectronics, and precision engineering, among others [6].

1.1.3.1 LBMM for Biomedical Applications

Silicon is the most commonly used material for biomedical devices because it is inexpensive, readily available, biologically inert, and easily integrated with silicon-based electronics. Silicon substrates may be used to fabricate templates for tissue substitutes as well as for micro fluidic channel fabrication [8]. In one case study reported by [8] pulsed argon fluoride laser (ArF) was used to machine micro channels on Silicon, Si (111). The study conducted by [8] showed an proportionate relationship between the aperture size and the ablation depth of the channel. Researchers, examined the human cell growth on micromachined channels as well as on the non-ablated Si. They found that Laser micromachining enhances the controlled alignment of the cells as well as the integration of the cells with the electronic devices [8].

In a study, conducted by [9] carried out an investigation into the use of femtosecond laser ablation for fabricating microfluidic channels on dielectric materials such as soda-lime glass and fused quartz, in combination with a polymer coating to modify surface roughness. We employed a systematic approach to determine the optimal range of ablation process parameters. The researchers [9] observed that increasing the pulse energy resulted in a higher rate of ablation; however, it also led to the generation of rougher surfaces due to elevated fluid forces and plasma pressure. The pulse energy was varied from 2.0 to 3.75 µJ, resulting in surface roughness measurements ranging from 295 to 731 nm RMS. By decreasing the pulse energy to 667 nJ, the surface roughness was improved to approximately 100 nm. Additionally, the application of HEMA coating effectively reduced the surface roughness to around 10 nm, irrespective of the initial surface roughness condition. Furthermore, they observed [9] that increasing the overlap at a given pulse energy decreased the surface roughness, as it improved the uniformity of laser fluence. Additionally, researchers [9] found that P-polarization of the laser beam was more effective than S-polarization in ablating the channels. While lower laser intensity or higher overlap can enhance surface roughness, there is a trade-off in terms of productivity and ablation rate. In practical engineering applications, the HEMA coating technique offers a high-quality surface roughness profile of approximately 10 nm without compromising productivity. Researchers further concluded [9] that there were no discernible changes in the chemical composition of the raw material after ablation; however, further investigation into the chemical composition was recommended for future studies [9].

Titanium-based alloy products produced using Selective Laser Melting (SLM) have gained significant popularity in biomedical applications due to their excellent biocompatibility and mechanical properties. To enhance the quality of Ti-6Al-4V parts fabricated through SLM and assist manufacturing engineers in selecting optimal process parameters, an optimization methodology based on an artificial neural network was developed by [10]. This methodology established [10] relationships between four key process parameters (laser power, laser scanning speed, layer thickness, and hatch distance) and two target properties of the fabricated parts (density ratio and surface roughness). A supervised learning deep neural network employing the backpropagation algorithm was utilized to optimize input parameters based on a given set of quality part outputs. Several techniques were implemented to address undesirable issues encountered during neural network training, thereby improving the accuracy of the model. The model demonstrated excellent performance, achieving an R-value of 99% for both density ratio and surface roughness. Subsequently, a selection system was constructed, enabling users to choose the optimal process parameters for fabricated products that meet specific user requirements. Experiments conducted using the recommended optimal process parameters from the optimization system strongly validated its reliability, as the resulting part qualities closely matched the user-defined criteria, with errors ranging from only 0.9 to 4.4% at maximum. Finally, a user-friendly graphical interface was developed to facilitate the selection of optimum process parameters for desired density ratios and surface roughness values [10].

Previous other studies have also documented diverse biomedical applications of the LBMM process [11]. For instance, Pereira et al. [12] investigated the potential of laser ablation combined with thermal treatment as a method to enhance the hydrophilicity of ceramic implants, thereby improving surface wettability [12]. Cunha et al. [13] demonstrated that utilizing the ultrafast LST technique on Ti-6Al-4V implants enhances surface wettability and influences the behavior of human mesenchymal stem cells (hMSCs) by affecting cytoskeletal morphology, distribution and area of focal adhesion proteins (FAPs), and proliferation [13]. Stango et al. [14] conducted a study involving hydroxyapatite (HAP) coating on laser-textured surfaces of 316LSS and Ti-6Al-4V implants, revealing that the laser-textured surface exhibits enhanced corrosion resistance and is suitable for biomedical applications [14]. Additionally, Yu et al. generated a microtexture on titanium surfaces and concluded that the structured texture promotes cell adhesion and plays a crucial role in contact guidance [15].

1.1.3.2   LBMM for Photonics Applications

Femtosecond laser (fs) is the main source for the LBMM applications in the field of photonics. Femtosecond laser micromachining of transparent materials offers distinct advantages over alternative techniques for fabricating photonic devices [16].

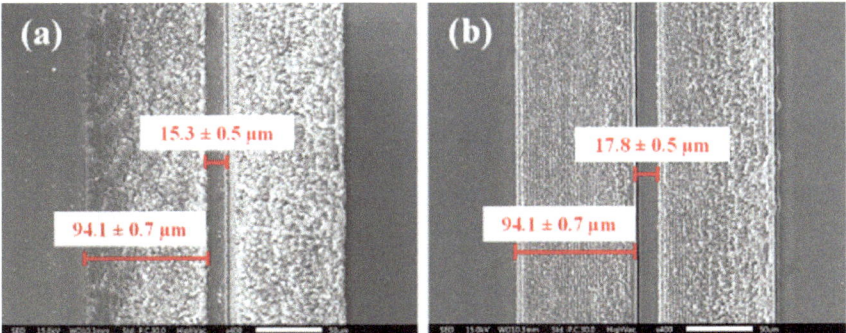

Fig. 1.4 Wave guide fabricated using femtosecond-based LBMM process [17]

Firstly, the nonlinear absorption characteristics confine any induced changes to the focal volume. This spatial confinement, when combined with laser beam scanning or sample translation, allows for the creation of intricate three-dimensional structures. Secondly, the absorption process is unaffected by the material, enabling the fabrication of optical devices on compound substrates composed of different materials. Thirdly, femtosecond laser micromachining facilitates the production of an "optical motherboard," where interconnects can be fabricated separately and later bonded to a single transparent substrate, either before or after the bonding of multiple photonic devices [16].

Lis et al. [17] presented the effectiveness of femtosecond laser micromachining (FLM) as a favorable method for microfabrication on multilayer structures compared to conventional techniques such as electron beam lithography and focused ion beam milling. They utilized the optimized FLM technique to fabricate one-dimensional photonic crystal (1DPhC) channel waveguides (comprises of SiO_2 and TiO_2) with widths of 15.3 ± 0.5 μm and 17.8 ± 0.5 μm. The waveguiding properties of these structures were then investigated using the end-fire coupling geometry. Figure 1.4 demonstrates the field emission scanning electron microscopic (FESEM) image of the LBMMed structure achieved by the researchers [17].

Federico Sala et al. [18] investigated the effect thermal annealing for the FLM process of glass substrate. They observed that introduction of the thermal annealing process significantly reduces the surface roughness produced by the FLM process. This reduction of the surface roughness enhances the optical performance of the FLMed features as investigated in [18].

1.1.3.3 LBMM for Precision Machining of Metals

Laser beam micromachining has been in use for varieties of metallic materials and their alloys which includes steel, copper, aluminum, titanium, nickel, etc. and their alloys [19]. Kaselouris et al. [20] investigated the heat affected zone as induced by the LBMM processing of metal substrate. They investigated various materials

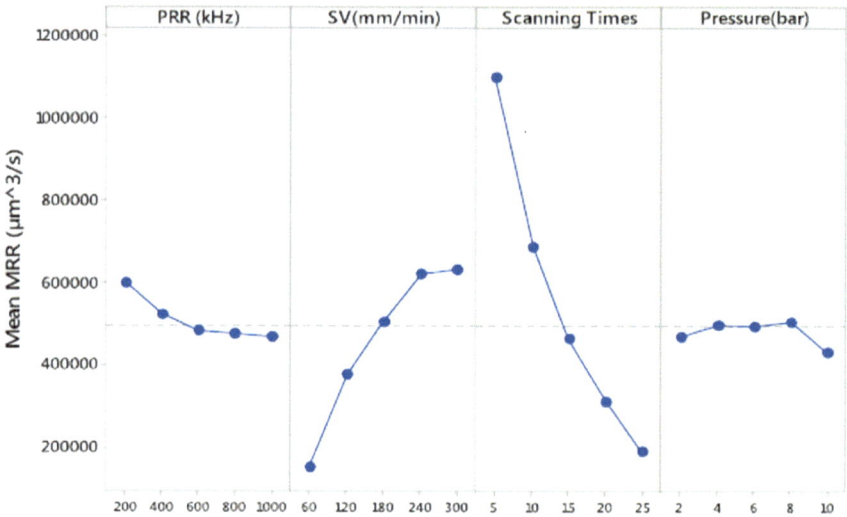

Fig. 1.5 Effect of various laser parameters on the MRR [22]

including AISI H13 steel, AISI 1045 steel and Ti6Al4V for this study that includes both finite element-based simulation study and the experimental validation of the simulation. The temperature distribution of the laser irradiated surface was found to be in good agreement with the simulation study. Balachninaite et al. [21] on the other hand investigated femtosecond laser micromachining of steel and copper substrate. Sahu et al. [22] studied the LBMM process for microchannel fabrication on titanium using nanosecond laser source. In their study [22] Sahu et al. [22] observed that the material removal rate (MRR) has a strong correlation with scanning velocity (SV) and scanning time as shown in Fig. 1.5. On the contrary Fig. 1.6 demonstrates that the average surface roughness of the LBMMed channels has a strong correlation with all the laser parameters including pulse repetition rate (PRR), SV, scanning time and air pressure [22]. Holder et al. [23] developed an analytical model to predict the LBMMed channels depth on titanium alloys which was found to be in good agreement with the experimental findings as shown in Fig. 1.7.

1.1.3.4   LBMM for Miscellaneous Use

The versatility LBMM has made this process a widely accepted method for micro-machining of different types of other materials also. Wang et al. [24] developed a MEMS-based piezoresistive pressure sensor where researchers used the LBMM process for fabrication instead of the dry etching process. Zin et al. [25] also investigated the LBMM for the MEMS fabrication process. They [25] were able to achieve LBMMed microchannel dimensions in the range of < 20 μm of depth with approximately ~ 2 μm of width (an example of the produced microchannel is shown in

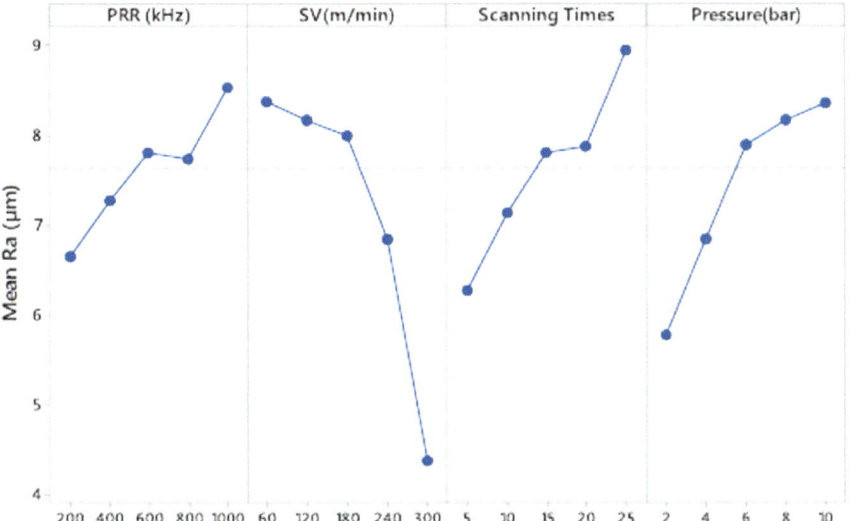

Fig. 1.6 Effect of the laser parameters on Ra [22]

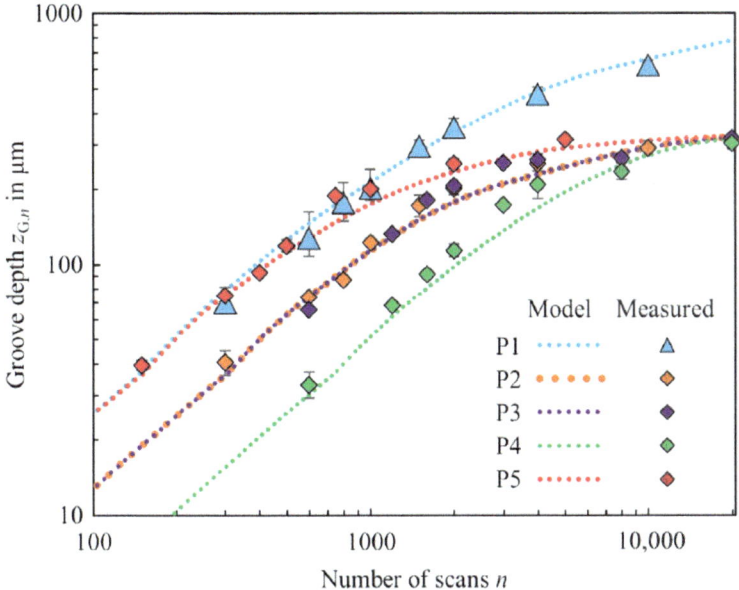

Fig. 1.7 Calculated groove depth (dotted lines, "Model") and measured groove depth (data points) for various laser parameters setting as described in [23]

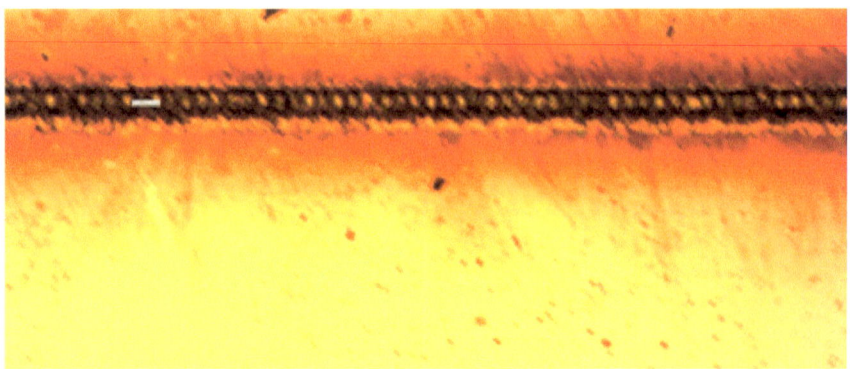

Fig. 1.8 LBMMed microchannels as produced by [25]

Fig. 1.8). The use of LBMM for the MEMS fabrication process has also been reported by other researchers [26, 27].

Silicon (Si) is another widely used engineering material in the field of electronics. Zhou et al. [28] explored the effects of ultrasonic power and water layer thickness on the new ultrasonic-assisted water confined laser micromachining method, specifically with a 1064 nm nanosecond pulse laser on silicon.

Figure 1.9 [28] shows the overall experimental setup as designed by Zhou et al. The water layer thickness h_w and ultrasonic power were varied in several levels. It was observed that both the ultrasonic power and water thickness had a positive impact on the depth of the machined groove by the LBMM process as shown in Fig. 1.10. The increase in the groove depth was observed due to the increased shock waves as produced by the applied ultrasonic transducer as explained by the researchers [28]. Zhao et al. [29] used FSL deep etching of Silicon Carbide to fabricate diaphragm of pressure sensor that can withstand high temperature. The researchers [29] developed an 80 µm-thick membrane as sensitive diaphragm for the pressure sensor. Micro-machining accuracy using the FSL was fond to be excellent with a reported error range less than 5% [29]. Other researchers also reported LBMM process for Si and Si wafers [30, 31].

1.2   Micro-Electrodischarge Micromachining (Micro-EDM/ µEDM)

Micro-Electrical Discharge Machining (Micro-EDM) is a precision manufacturing technique known for its ability to fabricate intricate components at micro scales. At its core, micro-EDM relies on electrothermal erosion, employing controlled and repeated electrical discharges to erode material from a workpiece. Unlike conventional machining methods, this process generates intense heat within microseconds,

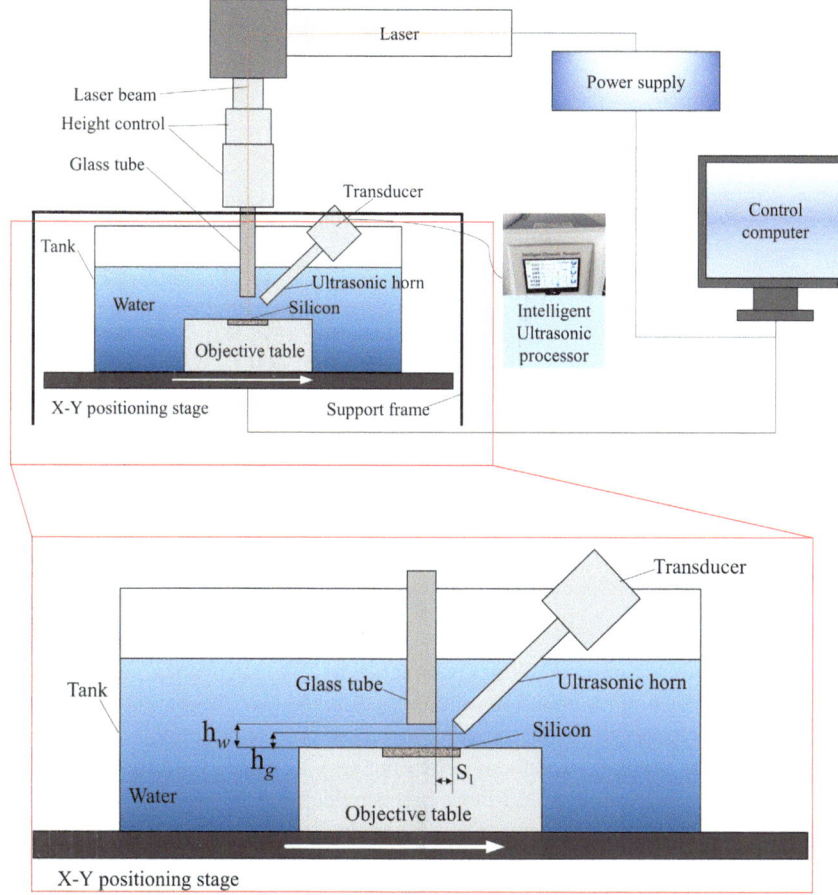

Fig. 1.9 Ultrasonic-assisted LBMM of Si [28]

vaporizing material and enabling the machining of electrically conductive materials regardless of their hardness.

1.2.1 History of Micro-EDM

Micro Electrodischarge Machining (micro-EDM), a non-traditional and precise machining process, has an illustrious history marked by continuous evolution to meet the advanced and intricate requirements of miniature manufacturing. It began as an extension of EDM, which was already well-regarded for its ability to machine complex shapes and hard materials that traditional machining couldn't achieve. In the early stages, micro-EDM was focused on the miniaturization of conventional

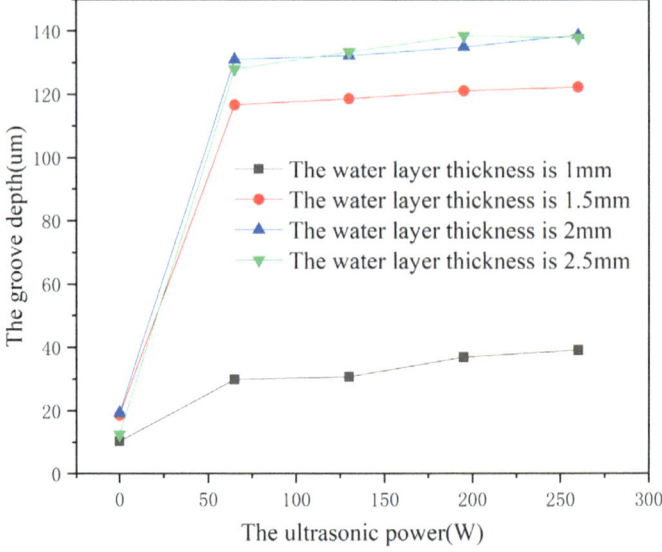

Fig. 1.10 Effect of ultrasonic power and water layer thickness on groove depth [28]

EDM processes to accommodate the manufacturing of smaller, intricate parts with high precision. Much of the initial research was directed toward refining the control of the discharge process to produce smaller and more precise features. This included developing power supplies capable of delivering very fine and controlled sparks, as well as the creation of smaller and more precise electrodes [32]. In the 1990s and early 2000s, advancements in electronics, materials science, and the need for minia-turized components in industries like aerospace, medical, and telecommunications, thrust micro-EDM into the spotlight. Researchers focused on improving various aspects of the process, such as the wear resistance of electrodes, surface finish, and machining accuracy, to make micro-EDM a more reliable and efficient process for the production of Micro Electromechanical Systems (MEMS) and other miniatur-ized devices [33]. Another significant advancement was the use of helical electrodes in micro-EDM, which provided improved flushing efficiency, better debris removal, and enhanced stability of the machining process compared to traditional straight electrodes. This innovation allowed for more complex geometries to be machined with better accuracy and has since become a standard in the field [34]. Through the years, continuous innovations such as machine intelligence, multi-axis control, and adaptive control systems were integrated into micro-EDM machines, making the process more versatile and capable of producing components with intricate details and high-aspect ratios. These developments have pushed the boundaries of what can be machined, leading to the manufacturing of components with complex geometries that were previously impossible to achieve. Today, micro-EDM is revered for its unmatched capability in machining intricate parts with high precision and accuracy, and it stands as a cornerstone in the field of precision manufacturing. Its history is an evidence to the initiative of engineers and researchers who have continually refined and perfected this process to meet the ever-increasing demands of modern technology and manufacturing.

1.2.2   Principle of Micro-EDM

The foundational concept of micro Electrodischarge Machining (micro-EDM) revolves around the precise removal of material from a conductive workpiece through a controlled process of electrical discharges. The workpiece is submerged in a tank containing dielectric fluid, and the material removal occurs in the vicinity of a sacrificial tool electrode, which is electrically charged by a high voltage power source as shown in Fig. 1.11 [35]. The tool electrode and workpiece are not in physical contact; rather, the high voltage generates an electric field that leads to the breakdown of the dielectric fluid, forming a plasma channel between the electrode and the workpiece. When the high-frequency pulsed electrical current is applied through either a transistor-based or an RC-based circuit, an intense localized heating occurs due to the electrical discharges, which results in the melting and vaporization of the material from the workpiece, reaching interfacial temperatures up to 20,000 °C [36]. This process is repeated in a meticulously controlled manner, facilitating the precise removal of material based on the set geometry and specifications [37, 38].

Material removal in micro-EDM is a complex process and is influenced by various factors such as discharge energy, pulse duration, electrode material, the concentration and types of dielectric fluid, and workpiece material composition. The interaction of these factors plays a critical role in defining the material removal rate (MRR), surface finish, and the geometrical accuracy of the machined part [39, 40]. The MRR has been the focal point of numerous researchers, as it is one significant aspect that determines the efficacy of the entire process [39, 41].

Electrode wear, albeit minor in comparison to material removal from the workpiece, is an integral aspect of the process that requires careful attention. It affects the overall precision of the micro-EDM process and has been subjected to an array of studies to minimize it further and enhance the electrode's lifespan and the accuracy of the machined features [42].

In summary it can be stated that micro-EDM is a sophisticated machining process that boasts high precision in microfabrication, and despite its relatively lower productivity in terms of MRR, it remains invaluable in manufacturing complex shapes that are challenging for conventional machining processes. Research in this area continues to optimize the process parameters and innovate new methods to improve efficiency and precision, all while understanding the in-depth dynamics of material removal at micro levels.

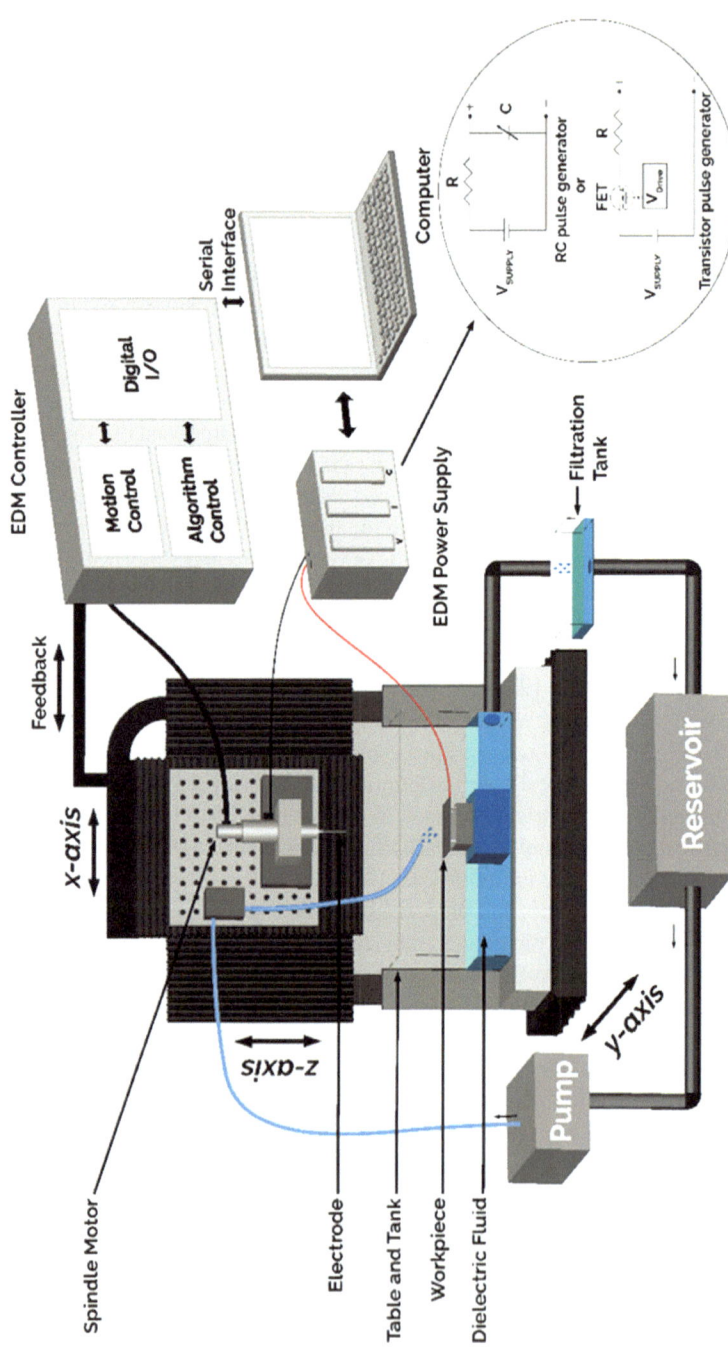

Fig. 1.11 Principle of micro-EDM process using RC and transistor-based pulse generator [35]

1.2.3 Application of Laser Beam Micromachining (LBMM)

Micro-Electrodischarge Machining (micro-EDM) stands out among precision manufacturing processes for its versatility and precision in creating microstructures and complex geometries on a variety of conductive materials. It has become a staple in the toolset of industries requiring high-precision components with intricate features [43]. Its capabilities are particularly beneficial in machining hard, brittle, or complex materials that pose challenges to conventional machining processes due to their high hardness or complex nature [44].

1.2.3.1 Micro-EDM for Manufacturing Precision Components

Lim et al. [45] conducted a study on micro-electrodischarge machining (micro-EDM) for fabricating high-aspect ratio microstructures highlighted the advantages of this method, including low set-up costs, high accuracy, and design flexibility. The research focused on two key processes: fabricating microelectrodes and EDM of the workpiece. Various operating parameters affecting micro-EDM, such as voltage, gap control, and circuit values, were explored to optimize the process. The study emphasized the importance of tool preparation, optimal machining conditions, and the development of an optical sensor for electrode measurement to ensure accuracy. Results from the study included insights into the fabrication of high-aspect ratio microstructures on an aluminum workpiece. The performance of the micro-EDM process was evaluated in terms of material removal rate (MRR), tool wear ratio (TWR), and machining stability. The influence of various operating parameters like voltage, gap control algorithm, and resistance and capacitance values in the spark control circuit were discussed. Additionally, experimental conditions for micro-EDM were outlined, including supply voltage variations, EDM circuit configurations, feed rates, spindle speeds, and dielectric coolant usage. The study also delved into the effects of different types of sacrificial electrodes on tool-electrode fabrication. Various electrode shapes fabricated using micro-EDM with different sacrificial electrodes were analyzed to understand their impact on shape accuracy and surface finish. The properties of the workpiece material (aluminum) and tool material (tungsten) were summarized to provide a comprehensive understanding of the materials used in the study.

Liu et al. [46] discusses the process capabilities of Micro-EDM, emphasizing its suitability for manufacturing accurate and complex three-dimensional microfeatures that are challenging to produce using conventional methods. The study explores newly developed technologies related to pulse generators and high-precision systems, essential for miniaturizing the EDM process. By examining pulse measurements, the correlation between discharge pulses and machine parameters is studied to understand the process capability. The research demonstrates optimized machining settings for different conditions, focusing on a ceramic composite Si3N4-TiN. Figure 1.12 shows the microhole machine by micro-EDM with an aspect ratio of 20. Figure 1.13

shows a microcompressor fabricated using the micro-EDM process as reported by [46].

In the study of [46] three machining regimes were investigated: roughing, semifinishing, and finishing, each requiring specific parameters for maximum material removal, low tool wear, and satisfactory surface roughness. The study shows that the advanced RC-based generator can produce very short discharge pulses, ensuring small unit material removal. The results indicate high feature accuracy achieved on stainless steel applications and excellent shape accuracy when machining the ceramic composite into complex three-dimensional structures. Overall, the study highlights

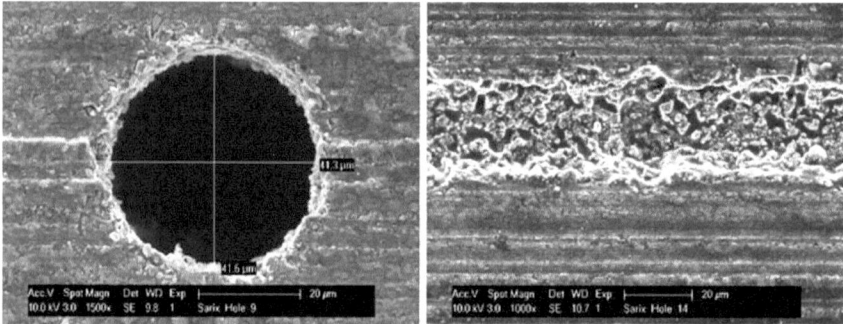

Fig. 1.12 Micro-EDM hole drilling on ceramic composite Si3N4-TiN with an aspect ratio of 20 [46]

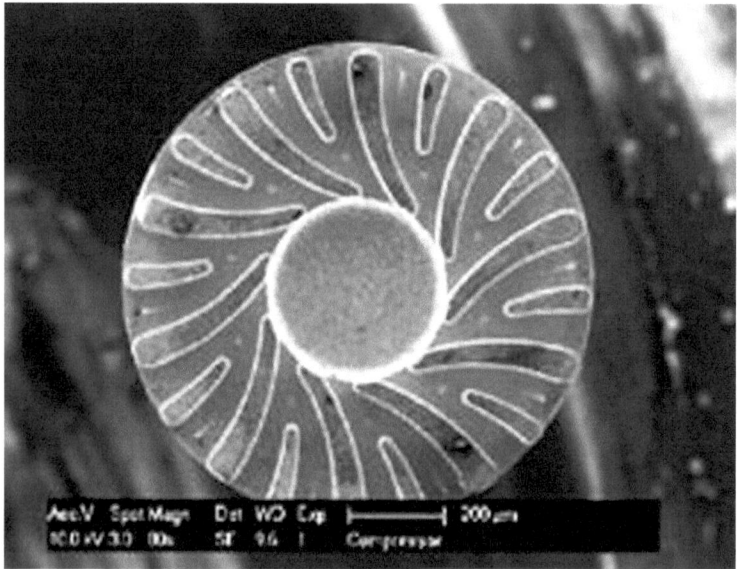

Fig. 1.13 Microcompressor fabricated by micro-EDM on 1 mm dia cylinder [46]

the advancements in Micro-EDM technology, showcasing its potential for producing intricate microfeatures with high precision and accuracy across various materials.

Dewangan et al. [47] conducted research focusing on optimizing the micro-EDM drilling parameters for Ti–6Al–4V alloy. They used an RC-based pulse generator as the EDM power supply. The capacitance was identified as the most significant parameter affecting Material Removal Rate (MRR), with voltage also playing a role. They used the Fuzzy-TOPSIS method for multi-criteria decision-making, leading to the identification of the best parameter settings. Optimal machining parameters for higher MRR and lower Overcut (OC) values were determined as capacitance of 1000 pF, voltage of 100 V, and tool rotation speed of 500 rpm. Analysis of variance (ANOVA) highlighted the significant impact of capacitance on both MRR and OC, while tool rotation speed had a lesser effect. The study emphasized the importance of optimizing parameters to enhance the quality and productivity of Ti–6Al–4V alloy machining processes.

Mouralova et al. [48] studied micro-EDM process for producing precision micro slots on copper foils. The study on optimizing the production of precision slots in copper foil using micro-EDM technology focused on investigating the influence of machine setting parameters, including pulse current, pulse-on time, and voltage, on the machining process responses. Through the design of experiment Box and Behnken Response Surface Design, the research aimed to achieve a precise slot measuring 5000×170 μm in a 125 μm thick copper foil. The results highlighted the significance of pulse-on time and voltage in affecting the erosion rate, emphasizing the need for statistical evaluation of the experimental design. The study showcased the advancements in micro-EDM technology for machining high-aspect ratio microstructures and its application in producing precision slots for optical devices, particularly in testing car headlights. Figure 1.14 shows various slots machined at different conditions on the copper foil. By utilizing statistical analysis, regression models, and optimization techniques, the research contributed to enhancing machining accuracy, efficiency, and the overall quality of microparts manufacturing.

1.2.3.2 Micro-EDM for Aerospace Applications

Das et al. [49] investigated microhole machining of aerospace materials like Inconnel 718 using micro-EDM. Super alloy Inconel 718 is highly sought after due to its widespread use in various engineering fields, thanks to its satisfactory machinability and exceptional physical and mechanical properties. This material garners attention from industry professionals and researchers worldwide. Micromachining of such materials has gained popularity in precision manufacturing industries like aerospace, automotive, and biomedical. A profound understanding of the machining process is crucial for working with challenging materials like Inconel 718, as various process parameters influence machining outcomes. EDM involves no direct contact between the electrode and workpiece, thereby eliminating mechanical forces and stresses. Additionally, it requires relatively low specific energy. The process delivers high accuracy in terms of surface roughness, typically reaching as low as 0.1 μm. This

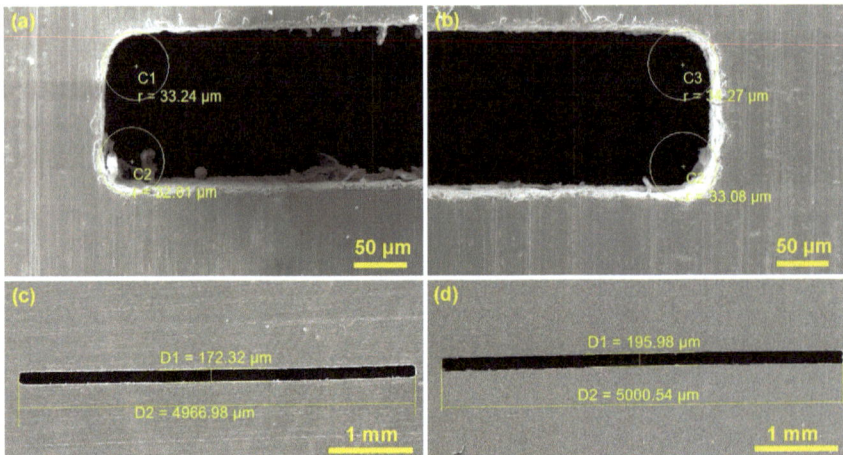

Fig. 1.14 Precision slot machined by micro-EDM on the copper foil. The subset **a** and **b** are showing the corner of the machined samples, **c** is the sample machined with the consistent width, and **d** is the sample machined with the consistent length [48]

research endeavors [49] to analyze the effects of key process variables—current, gap, voltage, pulse-on, and pulse-off time—on overcut. Experiments were conducted following a Box–Behnken design under Response Surface Methodology (RSM). Results indicate that current and pulse-on duration significantly impact overcut, while pulse-off duration and gap play crucial roles in minimizing it.

Titanium alloys, stainless steels, and high-speed steels are three types of high-performance alloys extremely popular for their use in the aerospace industry. Feng et al. [50] studied in detail the EDMing performance of these materials particularly useful for aerospace applications. The research explored the effects of various factors on surface morphology, including different materials, pulse power supplies, processing methods, and machining parameters like discharging energies, rotation speeds of the tool electrode, and gap voltages. It was observed that these factors significantly influenced the quality of surface morphology. The study highlighted material-specific observations such as the carbon increase rate in Ti-6Al-4V being the highest due to its properties, while SUS304 and SKH59 had different characteristics affecting their surface morphology during micro-EDM processing. High-speed rotation of the tool electrode was found to affect discharging channels, while different gap voltages influenced the completeness of the discharging process, ultimately affecting the surface quality. A comparison between two pulse power supplies showed that different circuit configurations led to varying discharge phenomena, impacting the resulting surface morphology. The study emphasized that pulse power supplies play a crucial role in determining the quality of machined surfaces.

1.2.3.3 Micro-EDM for Biomedical Applications

Jahan et al. [51] used the micro-EDM process for surface modification of Ti alloy (Ti-6Al-4V) for making it biocompatible to use for various biomedical applications. Micro-EDM was able to produce surface roughness below 100 microns, promoting cell growth around the implant. Moreover, the roughness of the machined surface could be controlled by managing crater sizes, showing a direct relationship between crater sizes and discharge energy. A porous and thick titanium oxide (TiO) layer was observed on the machined surface, facilitating bony ingrowth into the porous structure for implant fixation to bone. Importantly, no harmful toxic substances were found on the machined surface that could be detrimental to the human body. Finally, a slight increase in microhardness was noted after micro-EDM, which could be advantageous for wear resistance in biomedical orthodontic applications.

Shah et al. [52] investigated the fabrication of microrods on biomedical materials such as Ti-6Al-7Nb. The paper investigates the performance characteristics of micro-EDM dressing for fabricating microrods using Ti-6Al-7Nb as the workpiece material and brass as the tool material. The research focused on classifying pulses into normal, effective, and arcing types, with normal pulses contributing most to material removal. The findings revealed that the percentage of normal pulses increased with decreased discharge energy, while arcing was more prevalent at higher energy settings. Pulse frequency increased with lower energy settings due to the need for more contributing pulses. Specific energy, defined as the energy required to remove a unit volume of material, was found to decrease with higher discharge energy settings and along the length of the microrod due to reduced re-solidification of molten metal. Figure 1.15 shows the fabricated microrods as reported in [52]. The study provides valuable insights into optimizing micro-EDM dressing processes for fabricating microrods efficiently in applications like biomedical engineering.

1.2.3.4 Micro-EDM for MEMS Applications

The application of Micro-EDM in MEMS fabrication is significant due to its capability to produce intricate microscale features that are often required in MEMS devices. MEMS technology integrates mechanical elements, sensors, actuators, and electronics on a common silicon substrate through microfabrication technology, and it has a profound impact on various industrial and consumer products, enabling advancements in multiple domains. Heeren et al. [53] the microstructuring of silicon through electrodischarge machining (EDM) and its practical applications particularly in MEMS. It emphasizes the advantages of micro-EDM, such as cost-effectiveness, high precision, and the capability to fabricate intricate three-dimensional shapes that are challenging to achieve with other methods. The primary applications highlighted include rapid prototyping and small-batch production, showcasing structures like a ceramic force motor, micromirrors, an acceleration sensor, and a microspring produced using micro-EDM. Researchers [53] also reported the fabrication of elastic force motor as shown in Fig. 1.16. The research [53] elaborates on the process of

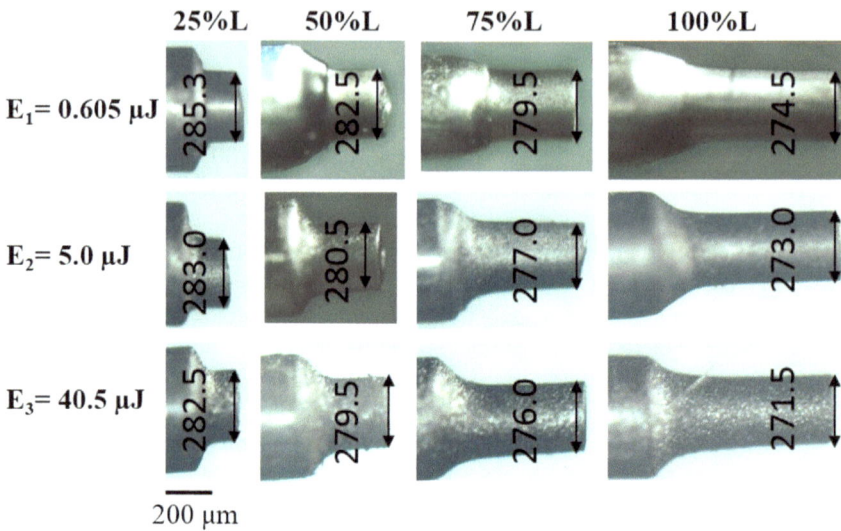

Fig. 1.15 Microrods with different aspect ratio fabricated at three different energy settings [52]

machining electrodes for micro-EDM, stressing the significance of preparing small electrodes directly on the EDM machine to ensure accuracy and prevent damage during transport. Modifications to the EDM machine for silicon micromachining are discussed, including the installation of a microgenerator, using deionized water as a dielectric, and incorporating a rotating clamping device for small electrodes to enhance accuracy. Additionally, details are provided on electrode machining techniques, electrode diameters ranging from 34 to 12 μm with high cylindricity and low roughness, and the influence of wear on sacrificial electrodes. The document also covers the successful machining of structures like mask membranes and 45° micromirrors in silicon, showcasing the feasibility and precision of micro-EDM in producing intricate silicon structures for MEMS applications.

Song et al. [54] investigated the application of Micro-EDM for fabricating silicon microstructures, focusing on surface roughness, microcracks, and dimensional control. It introduces micro-EDM as a solution to the limitations faced in producing three-dimensional structures on silicon due to crystal orientation constraints. The study emphasizes the influence of sparking energy on surface quality, microcracks, and the smallest producible micro beam thickness, highlighting that higher sparking energy leads to rougher surfaces and larger microcracks. The research identifies optimal voltage settings and capacitor values for achieving good surface quality, burr size, and dimensional accuracy in microstructure fabrication, with voltage being the most critical parameter. Additionally, the paper researches into the importance of sparking energy, capacitor values, and machining parameters in achieving desired surface roughness, dimensional accuracy, and machining efficiency in micro-EDM processes. The lowest surface roughness was found to be around 0.2 μm Ra. The discharge voltage of 100 V was the right setting in terms of surface roughness, bur

Fig. 1.16 Elastic force motor rotor [53]

size, dimension control, and cutting speed for the fabrication of 40–80 μm beams. The minimum thickness of the fabricated beam was 25 μm.

1.3 Summary

In the realm of micromachining, techniques such as Laser Beam Micromachining (LBMM) and Micro-Electrodischarge Machining (micro-EDM) have gained significant prominence due to their widespread application across diverse industries, including precision manufacturing, biomedical, automotive, aerospace, electronics, and MEMS (Micro-Electromechanical Systems) industry. While both LBMM and micro-EDM serve as indispensable tools in the manufacturing arsenal, each possesses distinct characteristics, accompanied by its own set of advantages and drawbacks. LBMM stands out for its remarkable speed and high production rate, making it a preferred choice for industries demanding rapid turnaround times. However, the trade-off for this efficiency often lies in compromised finished product quality when compared to micro-EDM counterparts. This discrepancy is vividly illustrated in Fig. 1.17, where (a) showcases an SEM image of a microhole machined via LBMM, while (b) depicts a micro-EDMed hole. Despite the significantly longer production time required for micro-EDM, the superior quality of the finished product is evident. Nevertheless, LBMM is not without its shortcomings. The process is prone to issues such as heat-affected zones (HAZ), recast layers, and tapperness, which can undermine the integrity and precision of the final product. To harness the strengths of both techniques and mitigate their respective weaknesses, there arises a compelling need to integrate these processes synergistically. The forthcoming chapters of this book delve into the strategies and methodologies for effectively integrating LBMM and micro-EDM, thereby leveraging the advantages of each to achieve an optimal balance

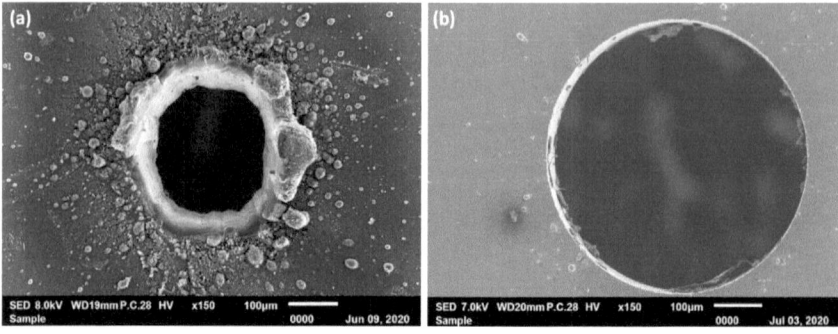

Fig. 1.17 **a** LBMMed hole, **b** micro-EDMed hole

between speed, precision, and quality in the production of micromachined components. This integration holds the promise of unlocking new frontiers in micromanufacturing, enabling the realization of intricately crafted and high-quality products across a spectrum of industries.

References

1. Masuzawa T (2000) State of the art of micromachining. CIRP Ann Manuf Technol 49(2):473–488. https://doi.org/10.1016/S0007-8506(07)63451-9
2. Rose M, Hogan H (2023) A history of the laser: 1960–2019. Photonics Media. Accessed: 03 Jul 2023 [Online]. Available: https://www.photonics.com/Articles/A_History_of_the_Laser_1960_-_2019/a42279
3. Parker S (2007) "Verifiable CPD paper: Introduction, history of lasers and laser light production. 13 Jan 2007. https://doi.org/10.1038/bdj.2006.113
4. Universal Laser System, History of Laser Technology. https://www.ulsinc.com/learn/history-of-lasers. Accessed 26 Jun 2023 [Online]. Available: https://www.ulsinc.com/learn/history-of-lasers
5. Klotzbach U, Lasagni AF, Panzner M, Franke V (2011) Laser micromachining. Adv Struct Mater 10:29–46. https://doi.org/10.1007/978-3-642-17782-8_2/FIGURES/16
6. Mishra S, Yadava V (2015) Laser Beam MicroMachining (LBMM)—a review. Opt Lasers Eng 73:89–122. https://doi.org/10.1016/j.optlaseng.2015.03.017
7. Rashid MAN, Saleh T, Noor WI, Ali MSM (2021) Effect of laser parameters on sequential laser beam micromachining and micro electro-discharge machining. Int J Adv Manuf Technol. https://doi.org/10.1007/s00170-021-06908-8
8. Miller PR, Aggarwal R, Doraiswamy A, Lin YJ, Lee Y-S, Narayan RJ (2009) Laser micromachining for biomedical applications. JOM 61(9):35–40. https://doi.org/10.1007/s11837-009-0130-7
9. Farson DF et al (2008) Femtosecond laser micromachining of dielectric materials for biomedical applications. J Micromech Microeng 18(3):035020. https://doi.org/10.1088/0960-1317/18/3/035020
10. Park HS, Nguyen DS, Le-Hong T, Van Tran X (2022) Machine learning-based optimization of process parameters in selective laser melting for biomedical applications. J Intell Manuf 33(6):1843–1858. https://doi.org/10.1007/s10845-021-01773-4

11. Shivakoti I, Kibria G, Cep R, Pradhan BB, Sharma A (2021) Laser surface texturing for biomedical applications: a review. Coatings 11(2):124. https://doi.org/10.3390/coatings11020124

12. Pereira RSF, Moura CG, Henriques B, Chevalier J, Silva FS, Fredel MC (2020) Influence of laser texturing on surface features, mechanical properties and low-temperature degradation behavior of 3Y-TZP. Ceram Int 46(3):3502–3512. https://doi.org/10.1016/j.ceramint.2019.10.065

13. Cunha A et al (2013) Ultrafast laser texturing of Ti-6Al-4V surfaces for biomedical applications. In: International congress on applications of lasers & electro-optics. Laser Institute of America, pp 910–918. https://doi.org/10.2351/1.5062989

14. Stango SAX, Karthick D, Swaroop S, Mudali UK, Vijayalakshmi U (2018) Development of hydroxyapatite coatings on laser textured 316 LSS and Ti-6Al-4V and its electrochemical behavior in SBF solution for orthopedic applications. Ceram Int 44(3):3149–3160. https://doi.org/10.1016/j.ceramint.2017.11.083

15. Yu Z, Yang G, Zhang W, Hu J (2018) Investigating the effect of picosecond laser texturing on microstructure and biofunctionalization of titanium alloy. J Mater Process Technol 255:129–136. https://doi.org/10.1016/j.jmatprotec.2017.12.009

16. Gattass RR, Mazur E (2008) Femtosecond laser micromachining in transparent materials. Nat Photonics 2(4):219–225. https://doi.org/10.1038/nphoton.2008.47

17. Sudha Maria Lis S, Rajasimha K, Debnath K, Krishna Chaitanya V, Bhaktha BNS (2022) Femtosecond laser micromachined one-dimensional photonic crystal channel waveguides. Opt Mater (Amst) 126:112114. https://doi.org/10.1016/j.optmat.2022.112114

18. Sala F, Paié P, Martínez Vázquez R, Osellame R, Bragheri F (2021) Effects of thermal annealing on femtosecond laser micromachined glass surfaces. Micromachines (Basel) 12(2):180. https://doi.org/10.3390/mi12020180

19. Raj D, Reddy BVR, Maity SR, Pandey KM (2019) ScienceDirect laser beam micromachining of metals: a review [Online]. Available: www.sciencedirect.com

20. Kaselouris E et al (2020) Analysis of the heat affected zone and surface roughness during laser micromachining of metals. In: Key engineering materials. Trans Tech Publications Ltd, pp 122–127. https://doi.org/10.4028/www.scientific.net/KEM.827.122

21. Balachninaitė O, Tamulienė V, Eičas L, Vaičaitis V (2021) Laser micromachining of steel and copper using femtosecond laser pulses in GHz burst mode. Results Phys 22. https://doi.org/10.1016/j.rinp.2021.103847

22. Sahu AK, Jha S (2020) Microchannel fabrication and metallurgical characterization on titanium by nanosecond fiber laser micromilling. Mater Manuf Processes 35(3):279–290. https://doi.org/10.1080/10426914.2020.1718702

23. Holder D, Weber R, Graf T (2022) Analytical model for the depth progress during laser micromachining of V-shaped grooves. Micromachines (Basel) 13(6):870. https://doi.org/10.3390/mi13060870

24. Wang L et al (2022) Development of laser-micromachined 4H-SiC MEMS piezoresistive pressure sensors for corrosive environments. IEEE Trans Electron Devices 69(4):2009–2014. https://doi.org/10.1109/TED.2022.3148702

25. Mohd Zin MZ, Felix EH, Wahab Y, Bakar MN (2020) Process development and characterization towards microstructural realization using laser micromachining for MEMS. SN Appl Sci 2(5):912. https://doi.org/10.1007/s42452-020-2715-2

26. Hausladen M, Buchner P, Schels A, Edler S, Bachmann M, Schreiner R (2023) An integrated field emission electron source on a chip fabricated by laser-micromachining and mems technology. In: 2023 IEEE 36th International Vacuum Nanoelectronics Conference (IVNC), IEEE, pp 115–116. https://doi.org/10.1109/IVNC57695.2023.10189001.

27. Oblov KY, Samotaev NN, Etrekova MO, Gorshkova AV (2019) Laser micromilling technology as a key for rapid ceramic MEMS devices. Phys At Nucl 82(11):1508–1512. https://doi.org/10.1134/S1063778819110152

28. Zhou J, Xu R, Jiao H, Bao J, Liu Q, Long Y (2020) Study on the mechanism of ultrasonic-assisted water confined laser micromachining of silicon. Opt Lasers Eng 132:106118. https://doi.org/10.1016/j.optlaseng.2020.106118

29. Zhao Y, Zhao Y-L, Wang L-K (2020) Application of femtosecond laser micromachining in silicon carbide deep etching for fabricating sensitive diaphragm of high temperature pressure sensor. Sens Actuators A Phys 309:112017. https://doi.org/10.1016/j.sna.2020.112017

30. Charee W, Qi H, Saetang V (2022) Underwater laser micromachining of silicon in pressurized environment. Int J Adv Manuf Technol. https://doi.org/10.1007/s00170-022-10120-7

31. Fang Z, Chen L, Guan Y, Zheng H (2020) Picosecond laser micromachining of silicon wafer: characterizations and electrical properties. Surf Rev Lett 27(05):1950142. https://doi.org/10.1142/S0218625X19501427

32. Gatzen HH, Klocke F, Kamenzky S, Traisigkhachol O (2008) Electroplated Cu micro electrode for application in micro electrostatic discharge machining (EDM). In: ECS meeting abstracts, vol MA2008-02, no 40, pp 2604–2604. https://doi.org/10.1149/MA2008-02/40/2604

33. Enciu C, Pârvu G, Ghiculescu L, Opran CG (2022) Application of micro electrical discharge machining and electrochemical machining in manufacturing of micro-electromechanical systems: a review. Macromol Symp 404(1). https://doi.org/10.1002/masy.202100449

34. Wang K, Zhang Q, Zhu G, Liu Q, Huang Y (2017) Experimental study on micro electrical discharge machining with helical electrode. Int J Adv Manuf Technol 93(5–8):2639–2645. https://doi.org/10.1007/s00170-017-0747-6

35. Hasan MM, Saleh T, Sophian A, Rahman MA, Huang T, Mohamed Ali MS (2023) Experimental modeling techniques in electrical discharge machining (EDM): a review. Int J Adv Manuf Technol 127(5–6):2125–2150. https://doi.org/10.1007/s00170-023-11603-x

36. McGeough JA (1988) Advanced methods of machining

37. Gostimirovic M, Radovanovic M, Madic M, Rodic D, Kulundzic N (2018) Inverse electro-thermal analysis of the material removal mechanism in electrical discharge machining. Int J Adv Manuf Technol 97(5–8):1861–1871. https://doi.org/10.1007/s00170-018-2074-y

38. Wong YS, Rahman M, Lim HS, Han H, Ravi N (2003) Investigation of micro-EDM material removal characteristics using single RC-pulse discharges. J Mater Process Technol 140(1–3):303–307. https://doi.org/10.1016/S0924-0136(03)00771-4

39. Majumder A (2012) Study of the effect of machining parameters on material removal rate and electrode wear during electric discharge machining of mild steel [Online]. Available: www.jestr.org

40. Sidpara AM, Malayath G (2019) Micro electro discharge machining. CRC Press/Taylor & Francis Group, Boca Raton. https://doi.org/10.1201/9780429464782

41. Hourmand M, Sarhan AAD, Sayuti M (2017) Characterizing the effects of micro electrical discharge machining parameters on material removal rate during micro EDM drilling of tungsten carbide (WC-Co). IOP Conf Ser Mater Sci Eng 241:012005. https://doi.org/10.1088/1757-899X/241/1/012005

42. Peng ZL, Li YN (2012) The deposition and removal process for micro machining based on electrical discharge. Adv Mat Res 472–475:2448–2451. https://doi.org/10.4028/www.scientific.net/AMR.472-475.2448

43. Jahan MP, Md. Ali Asad AB, Rahman M, Wong YS, Masaki T (2011) Micro-Electro Discharge Machining (μEDM). In: Micro-manufacturing, Wiley, pp 301–346. https://doi.org/10.1002/9781118010570.ch10

44. Chaitanya CRA, Wang N, Takahata K (2010) MEMS-based micro-electro-discharge machining (M^3 EDM) by electrostatic actuation of machining electrodes on the workpiece. J Microelectromechan Syst 19(3):690–699. https://doi.org/10.1109/JMEMS.2010.2047845

45. Lim HS, Wong YS, Rahman M, Edwin Lee MK (2003) A study on the machining of high-aspect ratio micro-structures using micro-EDM. J Mater Proces Technol 140:318–325. https://doi.org/10.1016/S0924-0136(03)00760-X

46. Liu K, Lauwers B, Reynaerts D (2010) Process capabilities of Micro-EDM and its applications. Int J Adv Manuf Technol 47(1–4):11–19. https://doi.org/10.1007/s00170-009-2056-1

47. Dewangan S, Kumar SD, Jha SK, Biswas CK (2020) Optimization of micro-EDM drilling parameters of Ti-6Al-4V alloy. Mater Today Proc 33:5481–5485. https://doi.org/10.1016/j.matpr.2020.03.307

48. Mouralova K, Bednar J, Benes L, Plichta T, Prokes T, Fries J (2022) Production of precision slots in copper foil using micro EDM. Sci Rep 12(1). https://doi.org/10.1038/s41598-022-089 57-9
49. Analysis on hole overcut during micro-EDM of Inconel 718-ScienceDirect [Online]. Available: https://www.sciencedirect.com/science/article/abs/pii/S2214785321075064
50. Feng W, Chu X, Hong Y, Zhang L (2018) Studies on the surface of high-performance alloys machined by micro-EDM. Mater Manuf Processes 33(6):616–625. https://doi.org/10.1080/104 26914.2017.1364758
51. Jahan MP, Alavi F, Kirwin R, Mahbub R (2018) Micro-EDM induced surface modification of titanium alloy for biocompatibility. Int J Mach Mach Mater 20(3):274–298. https://doi.org/10. 1504/IJMMM.2018.093548
52. Shah MS, Saha P (2021) Investigation on performance characteristics of micro-EDM dressing for the fabrication of micro-rod(s) on Ti-6Al-7Nb biomedical material. Mach Sci Technol 25(3):398–421. https://doi.org/10.1080/10910344.2020.1815050
53. Henri's Heeren P, Reynaerts D, Van Brussel H, Beuret C, Larsson O, Bertholds A (1997) Microstructuring of silicon by electro-discharge machining (EDM) part II: applications
54. Song X, Reynaerts D, Meeusen W, Van Brussel H, Investigation of micro-EDM for silicon microstructure fabrication [Online]. Available: http://proceedings.spiedigitallibrary.org/

Chapter 2
Laser-Micro-EDM-Based Hybrid Process

This chapter provides an overview of the Laser-Micro-EDM-Based Hybrid Process, focusing on its historical background and underlying principles. It begins with a retrospective examination of the origins of Laser-Micro-EDM technology, tracing its development and eventual integration with other machining techniques, leading to the concept of hybridization. Throughout this discussion, the challenges associated with this integration are explored, along with techniques devised to address them effectively. Additionally, the chapter investigates the influence of laser parameters on the Laser-Micro-EDM-Based Hybrid Process, examining their effects on both one-dimensional and three-dimensional machining. Lastly, the current state of Laser-Micro-EDM-Based Hybrid Processes is assessed, providing insights into their significance in contemporary manufacturing practices.

2.1 History and Principle of Laser-Micro-EDM-Based Process Hybridization

Researchers have shown significant interest in hybrid micromachining processes to improve microfabrication technology. Hybrid machining for micromachining has been subject to broad definitions among researchers. Aspinwall et al. [1] and Curtis et al. [2] define hybrid machining as the simultaneous utilization of all machining methods and the incorporation of two or more machining processes within a single machine. Hybrid machining was defined by Menzies et al. [3] as the integration of two or more machining processes in a synergistic manner to capitalize on the benefits of each. Laser beam micromachining (LBMM) and EDM have both been incorporated into hybrid machining technology. Moreover, electrochemical discharge machining (ECDM) is a hybrid machining concept that simultaneously facilitates electrochemical and electrodischarge machining. In contrast to pure EDM, ECDM exhibits the capacity to be employed in micromachining nonconductive materials [4]. Combining

© The Author(s), under exclusive license to Springer Nature Singapore Pte Ltd. 2025
T. Saleh et al., *Laser-MicroEDM Based Hybrid Micromachining*,
Manufacturing and Surface Engineering, https://doi.org/10.1007/978-981-97-8374-8_2

conventional grinding and EDM, electrical discharge abrasive grinding (EDAG) is an additional EDM-based hybrid machining technique [4]. Vibration-assisted EDM is another hybrid micromachining approach that has been the focus of study by numerous researchers [5].

Until recent years, numerous researchers documented investigations about laser-assisted machining procedures. Singh et al. [6] suggested laser-assisted mechanical micromachining for materials that are challenging to cut. The laser was employed to soften the hard material thermally during the mechanical interaction between the cutting tool and the workpiece. Sun et al. [7] proposed sequential hybrid micromachining combining LBMM and electrochemical micromachining (EMM). The subsequent EMM finishing procedure effectively eliminated the recast layer machining holes that were created during the LBMM process. Additionally, laser-assisted turning [8], laser-assisted milling [9], and laser-assisted waterjet machining [10] have been documented as laser-assisted machining methods.

Scholars have also documented sequential micromachining with laser-EDM technology [11–13]. Li et al. [11] presented a micromachining process based on LBMM-EDM to produce fuel nozzles. The suggested methodology contributed to drilling cost and time reductions of forty-two and seventy percent, respectively. Additionally, the production capacity was enhanced by 90% while maintaining the same level of hole quality as with pure EDM drilling. In addition, Li et al. [11] put forth a fixturing mechanism tailored to the fuel nozzle to ensure that the alignment of the hole remains within $\pm$ 20 μm. Kim et al. [12] also examined nanosecond pulsed laser and μEDM for micromachining of holes and microstructures. The reduction in machining time for microdrilling and micromilling processes exceeded fifty percent. Al-Ahmari et al. [13] examined LBMM-EDM-based microdrilling for shape memory alloy (SMA) based on Ni–Ti, which assisted in improving hole quality to that of pure LBMMed holes. A challenge faced in previous research concerning LBMM-μEDM sequential machining is the inadequate characterization of the alignment between the μEDM tool and the LBMMed pilot hole [11–13].

The principle behind Laser-Micro-EDM lies in the fusion of Laser Machining and traditional electrodischarge machining to achieve precise micromachining of materials. Initially, the process begins with laser machining, where a laser beam creates a pre-machined hole or cavity, laying the groundwork for subsequent EDM procedures. The EDM process then takes over, using electrical discharges to erode material and craft intricate features with exceptional precision.

This fusion of techniques allows for the production of intricate microstructures and components with outstanding accuracy and surface finish. Notable advantages of Laser-Micro-EDM include the faster processing enabled by using a laser for the initial machining step, as opposed to conventional EDM, which often requires pre-drilled holes. Additionally, Laser-Micro-EDM enables the machining of highly resilient or tough materials with intricate details, a challenge encountered with conventional machining methods.

The integration of laser and EDM technologies facilitates efficient micromachining, serving a variety of applications in industries such as medical, aerospace, and automotive, where precision and surface quality are of utmost importance.

2.2 Challenges and Mitigation Techniques

Efficient and accurate movement of workpieces between different machining stages is vital for effective manufacturing. However, challenges often arise that can hinder productivity and compromise the quality of the entire process. One such challenge is the time lost during the transfer of workpieces from a laser machine to an μEDM machine. This delay can disrupt production timelines. Additionally, misalignment errors during the transfer introduce inaccuracies, affecting the precision of the machining process. Simultaneously, there is a risk of compromising the integrity of the sample during this transition, potentially leading to dimensional issues and a decline in product quality. Addressing these challenges is crucial for optimizing manufacturing operations and ensuring a smooth flow of workpieces through different machining stages.

2.2.1 Time Loss During Workpiece Transferring from Laser to μEDM Machine

The challenge lies in the extended time required to move the workpiece from the laser machine to the μEDM machine. This transfer delay, involving the shift of the workpiece between different machining tools, results in a notable loss of valuable production time. This time lapse introduces an additional step that does not contribute value, consequently prolonging the overall manufacturing cycle. Addressing and minimizing this transfer time is crucial for improving operational efficiency and streamlining the manufacturing process.

2.2.2 Misalignment Error

Misalignment errors during the transfer from a laser machine to an μEDM machine refer to inaccuracies or deviations in the positioning and alignment of the workpiece as it transitions between these two machining tools. This misalignment can occur due to a variety of factors, including differences in the coordinate systems, variations in fixture setups, or inconsistencies in the alignment mechanisms of the two machines.

When a workpiece is not properly aligned during the transfer from the laser machine to the EDM machine, it can result in several adverse effects. Firstly, there may be a loss of precision in the subsequent μEDM process, leading to dimensional inaccuracies in the machined part. This misalignment error can compromise the intended geometrical features and specifications of the workpiece.

Moreover, misalignment errors can contribute to increased scrap rates, as poorly aligned workpieces may require rework or be deemed unusable. The overall efficiency of the manufacturing process is also impacted, as corrective measures to address misalignment consume additional time and resources.

2.2.3 Compromising the Sample Integrity While Transferring

Compromising sample integrity during the transition from a laser machine to an μEDM machine involves jeopardizing the overall quality, accuracy, and structural integrity of the workpiece. This issue arises when there are disruptions or inaccuracies in the transfer process that affect the intended characteristics of the sample.

The consequences of compromised sample integrity are significant. Firstly, there may be a negative impact on dimensional accuracy, with the final machined product deviating from the specified tolerances. This can lead to a decrease in product quality and reliability. Additionally, the structural integrity of the workpiece may be compromised, affecting its mechanical properties and overall functionality.

2.2.4 Mitigation Techniques

The tool setting process is a unique challenge for the LBMM-μEDM machining. To address the challenges and enhance the manufacturing process, an innovative solution has been implemented involving the utilization of a high-resolution On-Machine Measurement system (OMM). This system is designed to provide precise and detailed measurements directly on the machining equipment, allowing for real-time monitoring and adjustment.

By incorporating a high-resolution OMM, the manufacturing process gains the ability to capture intricate details and dimensions of the workpiece with exceptional accuracy. This technology enables continuous feedback during the machining operations, helping to identify and rectify any deviations or errors promptly. As a result, the risk of misalignment errors, dimensional inaccuracies, and compromised sample integrity is significantly reduced.

The high resolution of the OMM ensures that even subtle variations in the machining process can be detected, contributing to improved overall quality and precision. The real-time measurements also facilitate dynamic adjustments, allowing the system to adapt to changing conditions and maintain optimal performance throughout the manufacturing cycle.

2.3 Effect of Laser Parameters on the Laser-Micro-EDM-Based Hybrid Process

Rashid et al. [14] conducted experimental study to explore the influence of different laser parameters on the overall effectiveness of the LBMM-μEDM process. Initially, sub-millimeter holes were drilled using the LBMM method with a programmed diameter of 200 μm, and subsequent fine finishing was performed through the μEDM operation. The LBMM drilling was executed using a desktop fiber laser machine (ytterbium-doped) with a rated maximum power of 20 W (measured actual maximum power approximately ~ 16.7 W using a Gentec prento laser power meter), as depicted in Fig. 2.1. The laser employed is a pulsed laser with a wavelength of 1060 nm, and it has a focal length of 200 mm. Control over the laser beam for creating various patterns on the workpiece is facilitated by an X/Y galvanometer scanner (maximum scan angle ± 15° and resolution 12 μrad). Figure 2.1b illustrates the scanning strategy employed during the LBMM process for pilot hole creation, with a line spacing of 10 μm between each scan line. The entire laser system, including the galvanometer, can be managed through an integrated graphical user interface (GUI).

The LBMM system allows control over three parameters: laser power, pulse frequency, and laser scanning speed, while keeping the feature geometry fixed. Subsequent to the LBMM drilling, both the workpiece and the fixture were moved to the μEDM machine (DT110, Fig. 2.1c) from Mikrotools Pte. Ltd. The CNC programmable machine features a positional accuracy of ± 1 μm per 100 mm of travel length, with a programming resolution of 0.1 μm. The EDM power supply utilizes an RC pulse generator, and the system's stray capacitance, determined by monitoring single pulse discharge energy, was measured at 0.6 nF. The material under investigation is stainless steel (SS304) with a thickness of 0.2 mm. Table 2.1 provides details on the laser and μEDM parameters employed in this study. It is important to note that μEDM parameters remained constant, while the laser parameters experienced variation in four stages following a full factorial experimental design.

Initially, a square array of pilot holes was drilled with varying laser parameters in accordance with the experimental design. The time required to drill each hole using the LBMM process was recorded using a stopwatch, with a resolution of 1/100th of a second. In the subsequent phase, fine machining took place on the μEDM (DT110) machine. The overall sequential machining process is illustrated schematically in Fig. 2.2. For the precise fine machining using μEDM, a 500 μm tungsten tool was employed. To ensure the tungsten tool's positional accuracy above the laser-drilled holes, an on-machine measurement (OMM) system was devised. This system comprises a variable lens optical microscope equipped with a high-resolution digital camera. As depicted in Fig. 2.2, the OMM was installed adjacent to the μEDM spindle.

The procedure employed to accurately position the tungsten tool above the LBMMed holes is detailed as follows. Initially, the sample workpiece was securely affixed to the fixture, and arrays of LBMMed holes were machined. Subsequently,

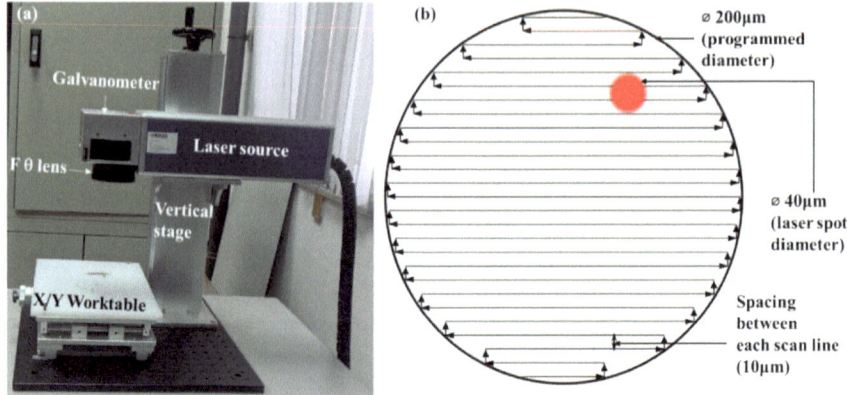

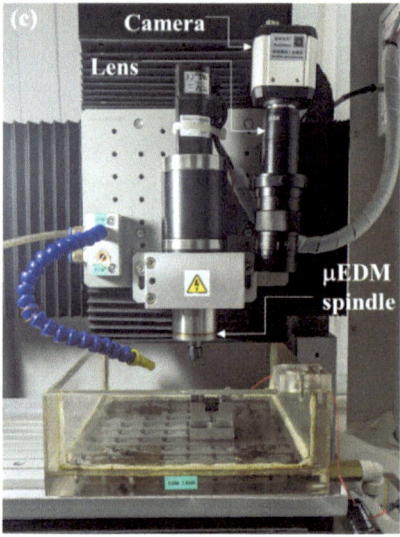

Fig. 2.1 **a** Fiber laser machining setup, **b** scanning strategy for the LBMM process, and **c** μEDM setup [14]

the entire fixture and workpiece were moved to the μEDM (DT110) machine, where a reference hole was machined in close proximity to the arrays of LBMMed holes using the μEDM process. The coordinates of the reference hole were obtained from the linear scale feedback of the μEDM machine. Following this, utilizing the camera (crosshair mark) and linear scale feedback, the deviation of X and Y coordinates for each LBMMed hole from the reference hole was measured. These deviations were then added to the coordinates of the reference hole. Consequently, the tungsten tool was precisely aligned directly over the center of the LBMMed holes, ensuring fine machining with acceptable accuracy (theoretical accuracy within 2 μm). To qualitatively assess the precision of the EDM tool's positioning method above the LBMMed

Table 2.1 Various parameters for laser and μEDM [14]

Parameter	Laser	uEDM
Average power (%)	15, 40, 65, 90	–
Measured average power (W)	6.4, 9.4, 12.5, 15.5	
Laser spot diameter (μm)	40	
Laser pulse duration (ns)	100	
Pulse frequency (kHz)	5, 10, 15, 20	–
Voltage (V)	–	80
Capacitor (nF)	–	1
Stray capacitance (nF)		0.6
Laser scanning speed (mm/s)/μEDM feed speed (μm/s)	50, 500, 950,1400	5
Set diameter of the hole for laser operation (μm)	200	–
Electrode diameter for μEDM (μm)	–	500
Loop count (Nos)	75	–
μEDM electrode rotational speed (RPM)	–	500
μEDM electrode material	–	Tungsten (W)

holes, the μEDM operation was conducted partially on one of the LBMMed holes, and an optical image was captured (Fig. 2.3). As depicted in Fig. 2.3, the developed OMM system facilitated accurate positioning of the μEDM tool over the pilot holes. Subsequently, the time required to drill each hole using μEDM was directly recorded from the machine, eliminating the need for a stopwatch.

The amplitude and frequency of the discharge current were measured utilizing a current probe (Tektronics: Tek CT1, sensitivity 5 mV/mA) attached to an oscilloscope (RIGOL). Throughout the EDM process, a potential short circuit, resulting from metal-to-metal contact, may occur without material removal [15]. In the event of a short circuit (detected by the DT110 machine's controller through monitoring the electrical characteristics of the μEDM process), a buzzer sound is activated, prompting the tool's forward motion to reverse until an open circuit is detected again. The occurrence of short circuits during each μEDM operation was quantified by counting the buzzer sounds. Notably, the DT110 machine can promptly detect a conductive surface upon contact in the X, Y, or Z direction.

To measure vertical tool wear, the tungsten tool was moved in the negative Z direction to detect a predefined reference surface. Upon detecting the surface, the Z coordinate was recorded both before and after machining each hole. The difference between these two Z coordinates was then calculated to determine the vertical tool wear. For each μEDM operation, the bottom portion of the tungsten tool underwent flattening using a process called reverse EDM, ensuring a consistent tool shape for machining all the holes [16].

The analysis of the machined samples was conducted using scanning electron microscopy (SEM: JSMIT100 InTouch Scope™). The characterization process occurred twice: first, after the LBMM operation, and finally, following the μEDM

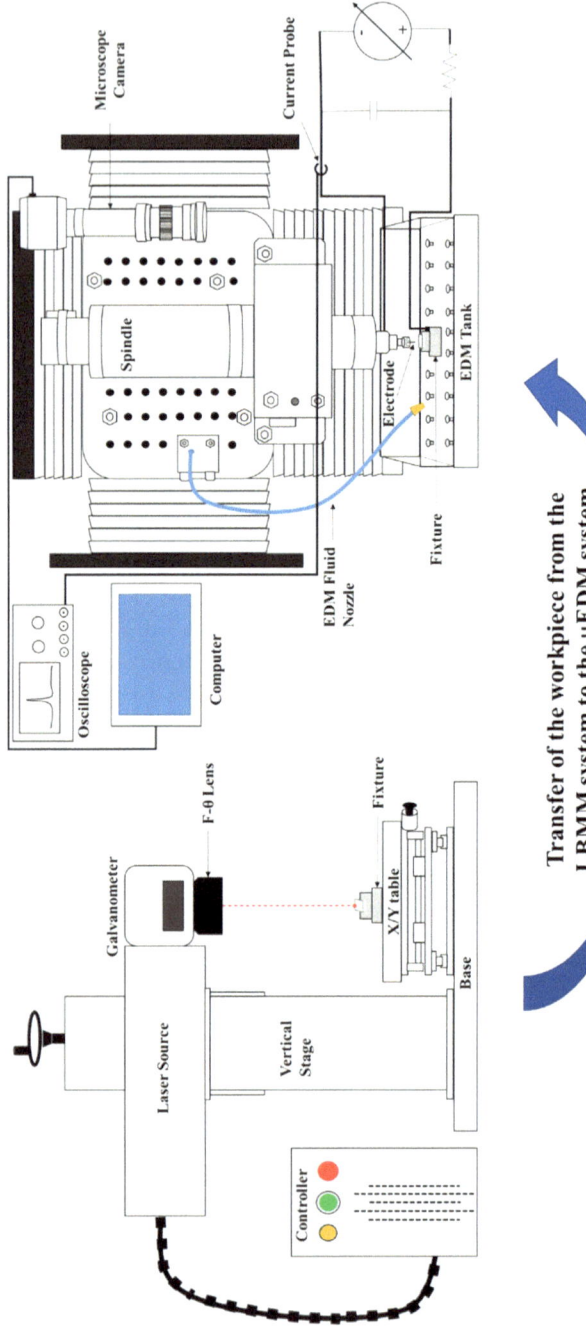

Fig. 2.2 Schematic illustration of the sequential micromachining process LBMM-µEDM [14]

Fig. 2.3 Effectiveness of the proposed centering method of the μEDM tool on the pre-machined LBMMed hole. Scale bar = 200 μm [14]

finishing. Subsequently, all SEM images were subjected to analysis using third-party image analysis software (ImageJ) [17] to quantify the entry and exit areas of both LBMMed and μEDMed holes. Measuring the volume of material removed during the LBMM process is a crucial parameter in this study. This measurement presents challenges as LBMMed holes lack a defined circular shape, and certain laser parameters used in this study were unable to penetrate a thru-hole on the workpiece.

Figure 2.4 illustrates an example of measurement using the image analysis software (ImageJ) [17]. After measuring the areas of the holes (entry area: Fig. 2.4a, c, exit/bottom area: Fig. 2.4b, d) and the slanted height (Fig. 2.4e), the removed volume by the LBMM process for each hole was calculated. The holes formed by the LBMM process resemble the shape of a cone with the top sliced off. If the entry area of the hole is B1, the exit area is B2, and the depth is h, then the formula for the volume (V) of material removed by LBMM is given by Eq. (2.1).

$$V = \frac{h}{3}(B_1 + \sqrt{B_1 B_2} + B_2) \tag{2.1}$$

The direct measurement of depth (h) for holes that were not fully penetrated was not feasible due to the unavailability of cross-sectional images for such holes. Therefore, a formula expressing h in terms of the slanted height, as represented by Eq. (2.2) in the derivation, was utilized for holes similar to those depicted in Fig. 2.4e. However, for fully penetrated holes, exemplified by Fig. 2.4a, b the thickness of the workpiece was considered as the value for *h*.

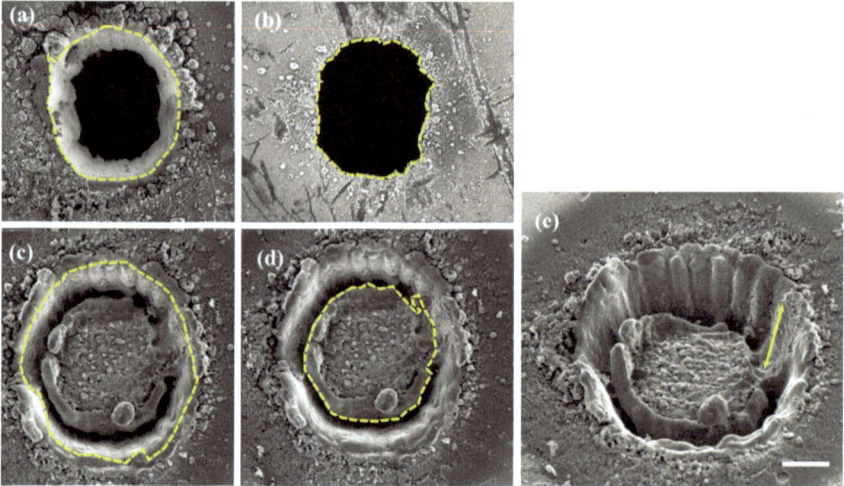

Fig. 2.4 Measurement of LBMMed holes entry, exit, slanted height, and the bottom area for calculating the volume removed by the LBMM process: **a** entry area of a thru-hole, **b** exit area of a thru-hole, **c** entry area of a blind hole, **d** bottom area of the blind hole, and **e** slanted height of the blind hole. Scale bar $= 100\,\mu m$ [14]

$$h = \sqrt{s^2 - \frac{B_1}{\pi} - \frac{B_2}{\pi} + 2\frac{\sqrt{B_1 B_2}}{\pi}} \qquad (2.2)$$

The method described above for quantifying the removed volume during the LBMM process proved sufficiently accurate in capturing the average trend of material removal through this process, accounting for variations in laser power and scanning speed.

Similarly, three-dimensional machining, which is μEDM milling, was performed on an identical workpiece. Loops, scanning speed, power, and pulse repetition rate were the parameters of the laser milling process under investigation.

2.3.1 Effect of Laser Parameters for One-Dimensional Machining

Figure 2.5 presents a comparative analysis of hole quality across three utilized processes: LBMM, μEDM, and LBMM-μEDM. The LBMM method excels in machining time, allowing holes to be drilled under 1 min. However, LBMM exhibits the poorest performance concerning hole quality, resulting in a thick recast layer and low circularity. In contrast, the quality of holes machined by μEDM and the LBMM-μEDM sequential process is nearly identical. Yet, the LBMM-μEDM sequential process demonstrates a significantly lower drilling time, approximately 2.6 times less than the pure µEDM method, as illustrated in Fig. 2.6. These findings align with previous research results. Figure 2.6 also indicates a substantially reduced occurrence of short circuits during the LBMM-µEDM process compared to standard µEDM (approximately 40 times less), suggesting greater stability in LBMM-µEDM compared to pure µEDM. Furthermore, vertical tool wear is nearly nine times lower in sequential micromachining. The improved stability and lower tool wear in LBMM-µEDM are attributed to the reduced material removal compared to pure µEDM.

Additionally, Fig. 2.7 displays scanning electron microscopic views of the holes, revealing that the LBMM process produces a pilot hole with a significant recast layer (similar to Fig. 2.5) and a non-uniform inner surface (a). This recast layer can be entirely removed through the subsequent µEDM-based finishing operation, resulting

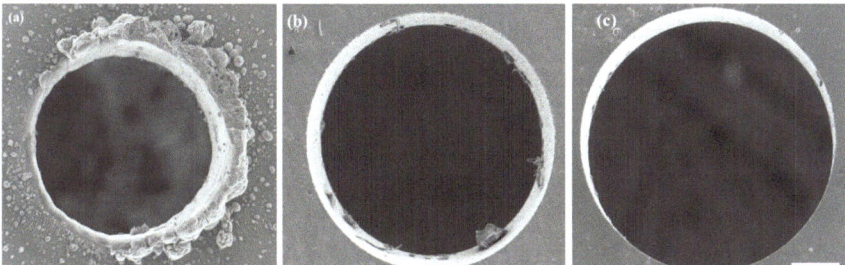

Fig. 2.5 Morphological comparison of the microholes machined by **a** pure LBMM process, **b** pure µEDM process, and **c** LBMM-µEDM-based sequential process. Laser processing was carried out at 15.5 W laser power, 50 mm/s scanning speed, and 20 kHz pulse frequency. Scale bar = 100 µm [14]

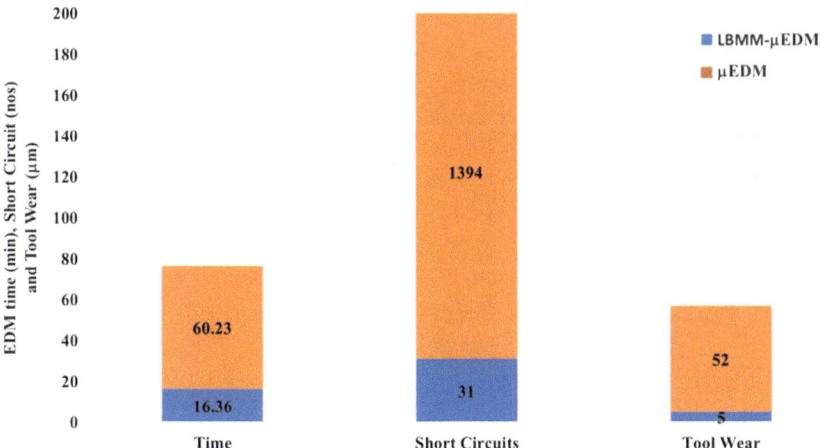

Fig. 2.6 Comparison of machining time, nos of short circuits, and tool wear between LBMM-µEDM and pure µEDM [14]

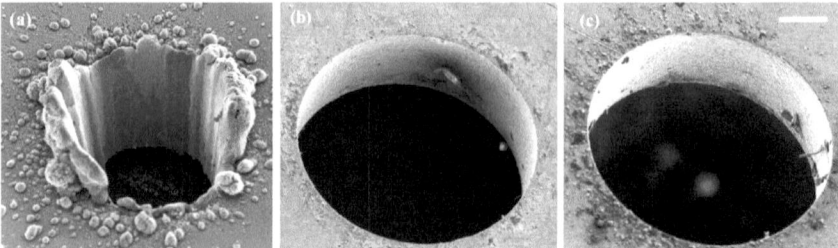

Fig. 2.7 Morphological study of the microholes **a** initially machined by the LBMM process, **b** fine-finished by the μEDM process, and **c** fully machined by the μEDM process. Scale bar = 100 μm [14]

in a hole with a uniform inner surface (b). Importantly, the quality of the LBMM-μEDM hole (b) matches that of the pure μEDM hole (c) regarding internal surface uniformity and minimized recast layer.

In the subsequent discussion, we will investigate the influence of different LBMM parameters on various performance indicators: machining time, stability and tool wear, and residual spatter zone in the context of LBMM-μEDM-based sequential micromachining. It is worth noting that, among the three examined laser parameters (pulse frequency, scanning speed, and power), pulse frequency did not significantly impact the performance of the sequential micromachining process.

2.3.1.1 Effect of Laser Parameters for One-Dimensional Machining

The study was carried out using an LBMM-μEDM-based sequential micromachining technique. The LBMMed holes were initially fine-finished using LBMM and refined using μEDM machining. The LBMM machining time was negligible in comparison to the μEDM finishing time. The variation in μEDM machining time for pilot holes drilled with different laser powers, scanning times, and pulse frequencies is illustrated in Fig. 2.8. The findings validate that pulse frequency does not significantly impact the μEDM machining time. It was noted that decreasing the scanning speed and increasing the laser power for pilot hole machining results in a shorter time required for fine finishing by EDM. On the contrary, the opposite impact is observed in holes machined with a fast-scanning speed in LBMM.

As the laser pulse frequency was found to have a random effect on the final machining time by μEDM, it was averaged up. Figure 2.9 illustrates the impact of scanning speed and laser power (averaging the frequency effect) on the ultimate μEDM time to provide a better quantitative understanding of the entire procedure. A scanning speed equal to or less than 200 mm/s in conjunction with a laser power ranging from 6.4 to 15.5 W during the LBMM operation will result in a final μEDM operation that may be concluded in thirty minutes, as seen in Fig. 2.9. Moreover, when the laser power is higher or equal to 13 W, even the most rapid scanning speed can

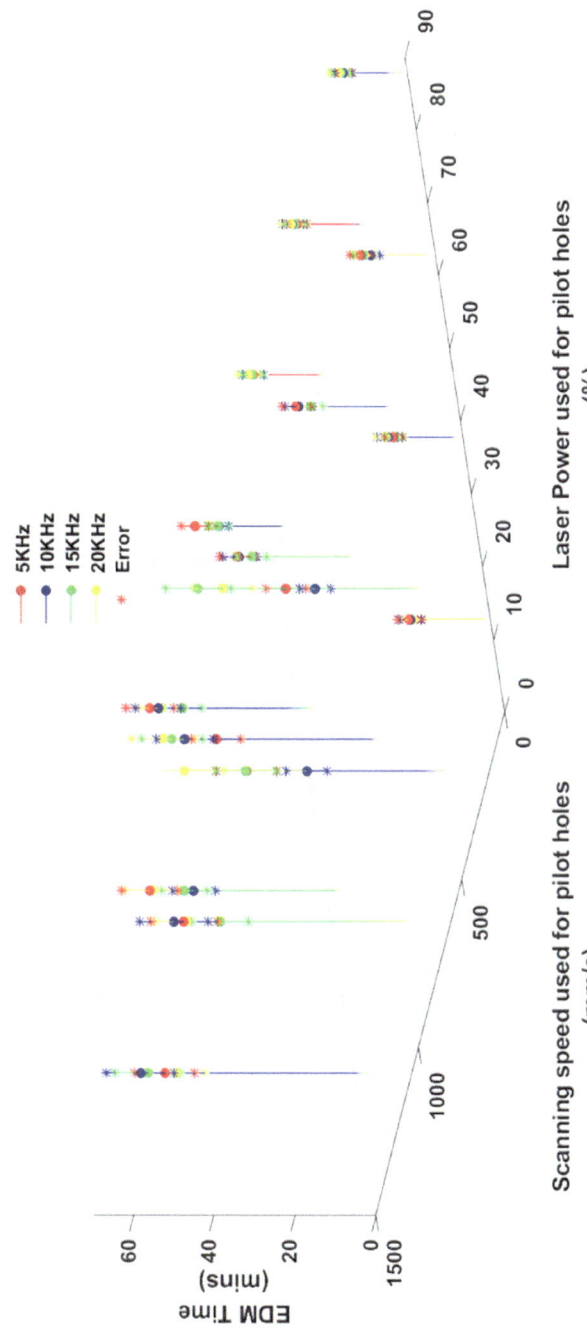

Fig. 2.8 Effect of the incident laser power, scanning speed, and pulse frequency (used for the pilot hole machining) on the μEDM processing time for the final finishing of the pilot holes. The error bar represents the machining uncertainty [14]

be utilized for LBMM machining. Nevertheless, the finishing process using μEDM may be completed in 30 min or less.

The average discharge frequency and discharge current, as determined during the μEDMing process of the LBMMed holes, are presented in Fig. 2.10. Furthermore, the data in Fig. 2.10a, b supports the conclusion that the machinability of LBMMed holes processed with higher laser power and lower scanning speed improves with increasing discharge frequency and discharge value [18]. This improvement can be attributed to the enhanced ease of machining. An analysis of the volume, entry, and exit areas of material extracted from the LBMMed holes during the LBMM operation can clarify the observation depicted in Figs. 2.8 and 2.9.

The impact of laser power and pulse frequency on the initial hole's entry area at various laser scanning speeds is seen in Fig. 2.11. Figure 2.11a–d shows that the entry area of the holes shows an increasing trend with the laser intensity. Nevertheless, no meaningful correlation exists between the laser's pulse frequency and entry area.

Additionally, the SEM pictures of the holes machined at various laser powers are shown in Fig. 2.12, which further supports the increasing trend of the incident laser power with the entrance area (Fig. 2.12a–d). However, Fig. 2.13 shows that an increase in laser scanning speed generally decreases the LBMMed holes' entry area. According to Negarestani et al. [19], the incident energy density in the LBMM process is inversely proportional to the scanning speed and proportional to the laser power. In our study, the entry area of the hole increased along with the incident energy as the laser intensity increased.

Likewise, lower energy density caused by faster scanning speed resulted in a smaller entry area of the holes. As seen in Fig. 2.14a–d, phenomena comparable to the entry area were noted concerning the LBMMed holes' exit area. As can be seen from the SEM pictures in Fig. 2.15a–d, it can be stated that the effect of scanning speed was found to be more dominating on the variation of the exit area compared to the entering area. Some holes were extensively pierced throughout the complete power range at a slow scanning speed of 50 mm/s. Laser power less than or equal to 9.4 W could not melt and evaporate the material sufficiently to produce LBMMed holes with a significant exit area since the scanning speed was similar to or greater than 500 mm/s.

Furthermore, Fig. 2.16 shows that as average laser power rises and scanning speed increases, the volume of material removed during the LBMM process decreases. The μEDM machining time trend seen in Figs. 2.8 and 2.9 results from the volume of LBMMed holes caused by variations in laser power and scanning speed. As previously indicated, during the LBMM process' pilot hole machining, a higher incident laser power and a slower scanning speed resulted in considerable material loss. Consequently, less material had to be removed by the machine during the μEDM operation, resulting in a faster processing time (Figs. 2.8 and 2.9).

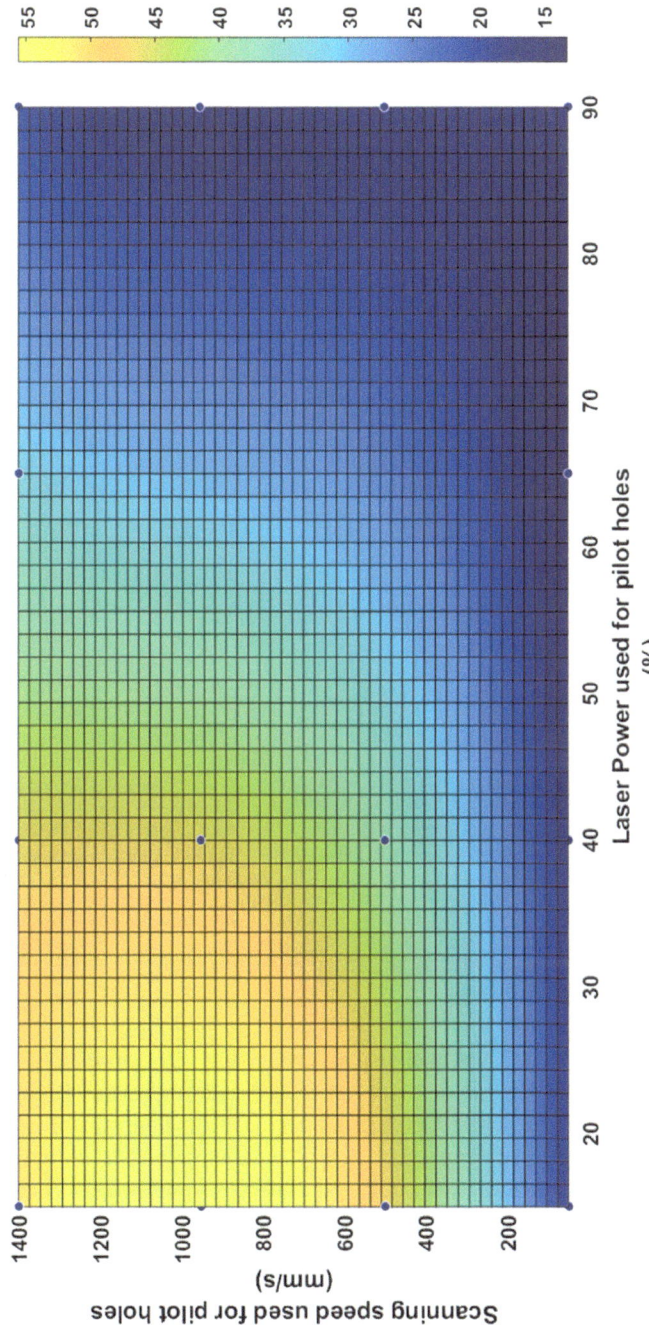

Fig. 2.9 Zone of faster and slower μEDM machining time as a function of laser incident power and laser scanning speed (used for pilot hole drilling using the LBMM process) [14]

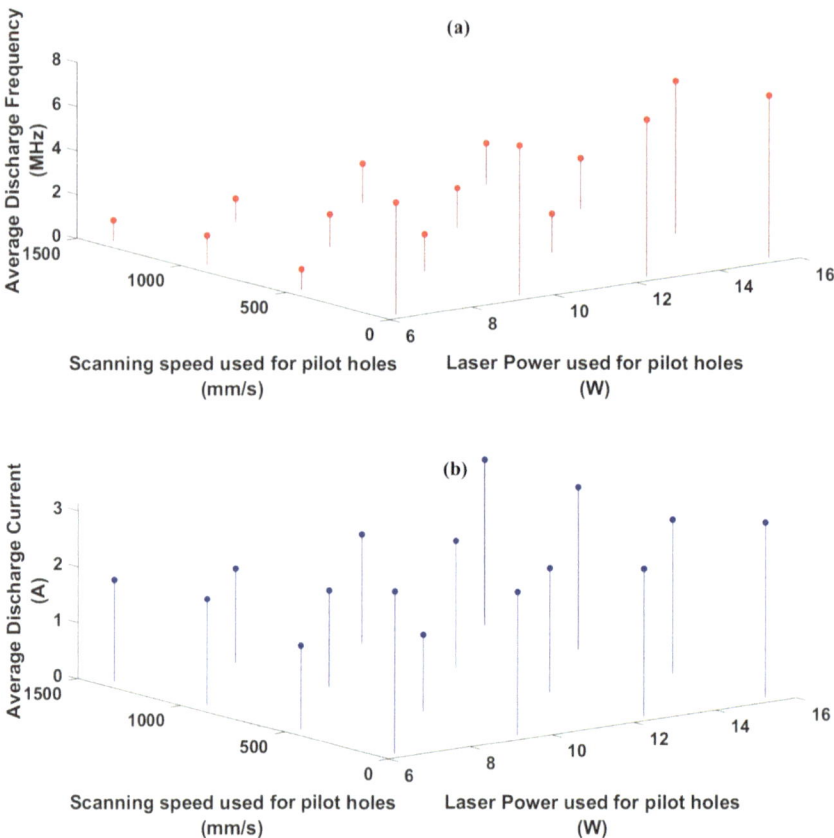

Fig. 2.10 Average discharge current and discharge frequency variation during the μEDM process as a function average laser power and scanning speed used for the pilot hole machining, **a** variation of the discharge current and **b** variation of the discharge frequency [14]

2.3.1.2 Study of MEDM Machining Stability and Tool Wear for Laser-MEDM Process

The stability of the μEDM process (specifically, the secondary operation of LBMM-μEDM machining) is shown by the frequency of short circuits that occur throughout the operation. If a short circuit is detected during the μEDM procedure on the DT110, the electrode's tool path will be reversed to rectify the temporary short circuit state. The EDM operation is believed to be unstable if an excessive number of short circuits are identified during a single machining operation. μEDM, being an electrothermal process, induces electrode wear due to the recurrent sparks.

It is evident from Fig. 2.17a–d that during LBMM-based rough drilling, the utilization of higher laser power and slower laser scanning speed reduces both tool wear and short circuit during EDM finishing. The observed phenomenon can be understood

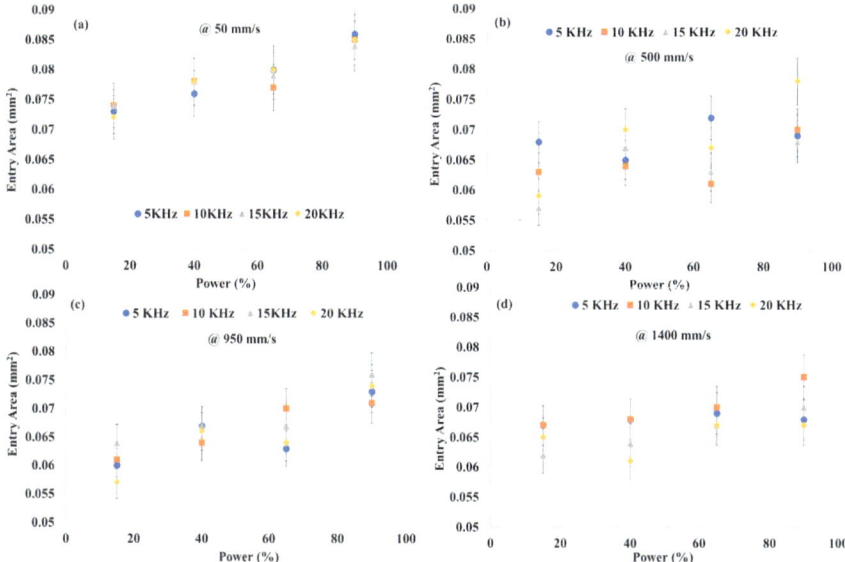

Fig. 2.11 Effect of the incident laser power and pulse frequency on the entry area of the pilot holes machined by the LBMM process. **a** The scanning speed was at 50 mm/s. **b** The scanning speed was at 500 mm/s. **c** The scanning speed was at 950 mm/s. **d** The scanning speed was at 1400 mm/s. The loop count for all cases was 75. The error bar shows the overall uncertainty [14]

by referring to Fig. 2.18, which illustrates a strong positive correlation (coefficients of 0.92 for tool wear and 0.99 for the number of short circuits during μEDM) with respect to the duration of the μEDM (Fig. 2.18a, b respectively). Consequently, longer interaction between the tool and workpiece during an enhanced EDM time will result in a greater risk of short circuits and tool wear.

It is validated in Fig. 2.19 that when employing a slower scanning speed and a higher laser power for LBMM drilling, the μEDMing time necessary for the precise finishing of the LBMMed holes is reduced. Consequently, this results in a decrease in tool wear and the occurrence of short circuits during the operation. When the LBMM process is executed at a scanning speed of 1500 mm/s and a power of 6.4 W, the tool wear and short circuit can reach values of around 58 μm and 1045 μm, respectively. These values are nearly equivalent to those observed in pure μEDM. However, when employing a power level of 15.5 W or greater across all scanning speeds, the tool wear and number of short circuits will be decreased to approximately 10 μm and 100, respectively. These figures represent a significant reduction compared to the pure μEDM (Fig. 2.6).

Furthermore, Fig. 2.19a, b demonstrates that the average rate of tool wear and the average rate of short circuits are both highly influenced by the laser input parameters used for pilot hole cutting, in addition to the total tool wear and total short circuits. Tool wear and short circuit rates can reach 1 μm/min and 20 nos/min, respectively, during the μEDM and LBMM processes when a high laser scanning

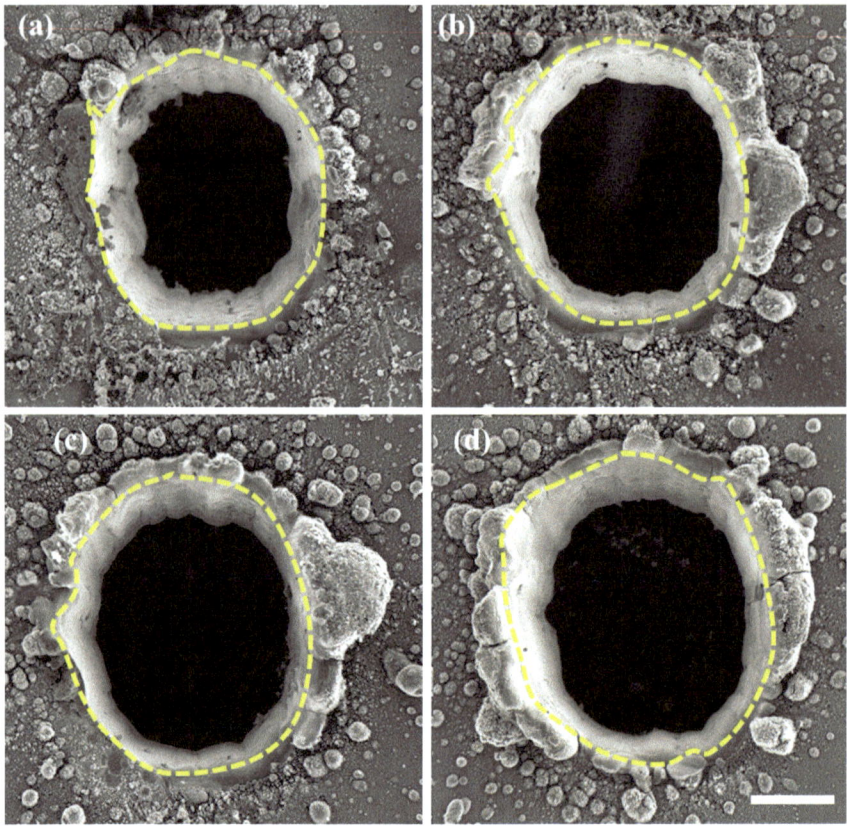

Fig. 2.12 SEM images of the entry area of the LBMMed holes machined by different incident laser power. **a** 6.4 W laser power, **b** 9.4 W laser power, **c** 12.5 W laser power, and **d** 15.5 W laser power. The loop count was 75, the scanning speed was 50 mm/s, and the laser pulse frequency was 5 kHz for all the holes. Scale bar = 100 μm. Black edge marker is used to highlight the entry area [14]

speed is employed with a low laser incident power. When the LBMM method is executed with low laser power and high scanning speed, the volumetric dimensions of the LBMMed pilot holes diminish (Fig. 2.16). Consequently, the tool must remove a greater quantity of material from the workpiece during μEDM, leading to an increase in the creation of debris. Consequently, the rate of short circuit and tool wear escalates. However, in the case of LBMMed holes, where a greater volume of material is removed, the μEDM operation requires the removal of less material. Consequently, we observed an increased frequency of sparking and a high discharge current during the process, as determined by monitoring the discharge current. As a result, both the frequency of short circuits and tool wear were reduced.

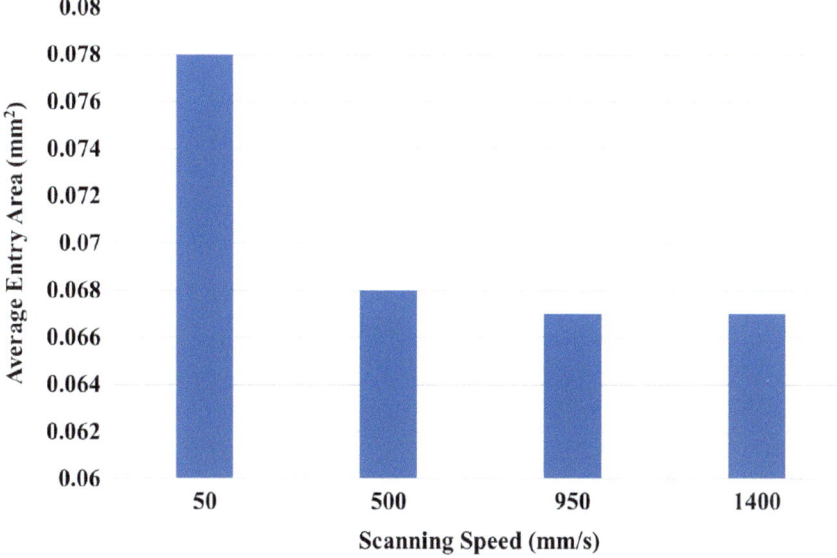

Fig. 2.13 Average effect of the laser scanning speed on the entry area of the pilot holes machined by the LBMM process. All the data for each scanning speed has been averaged up to plot this graph [14]

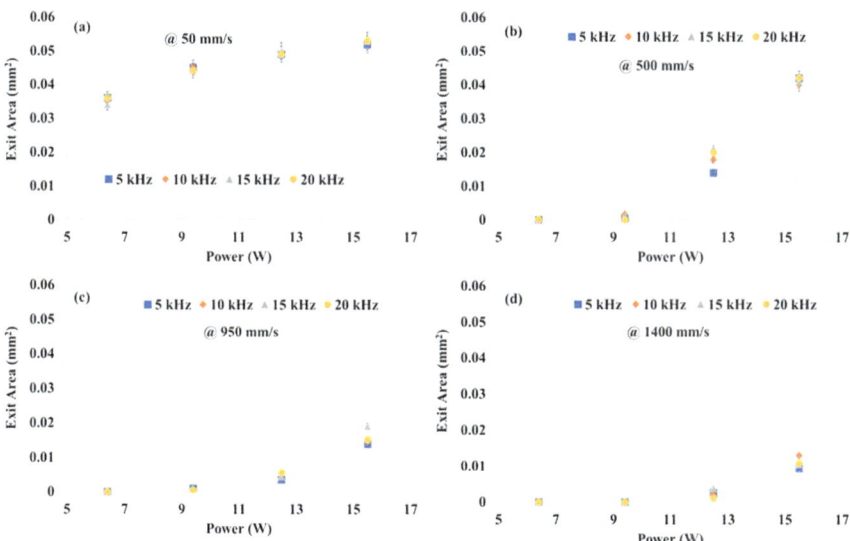

Fig. 2.14 Effect of the incident laser power and pulse frequency on the exit area of the pilot holes machined by the LBMM process. **a** The scanning speed was at 50 mm/s. **b** The scanning speed was at 500 mm/s. **c** The scanning speed was at 950 mm/s. **d** The scanning speed was at 1400 mm/s. The loop count for all cases was 75. The error bar shows the overall uncertainty [14]

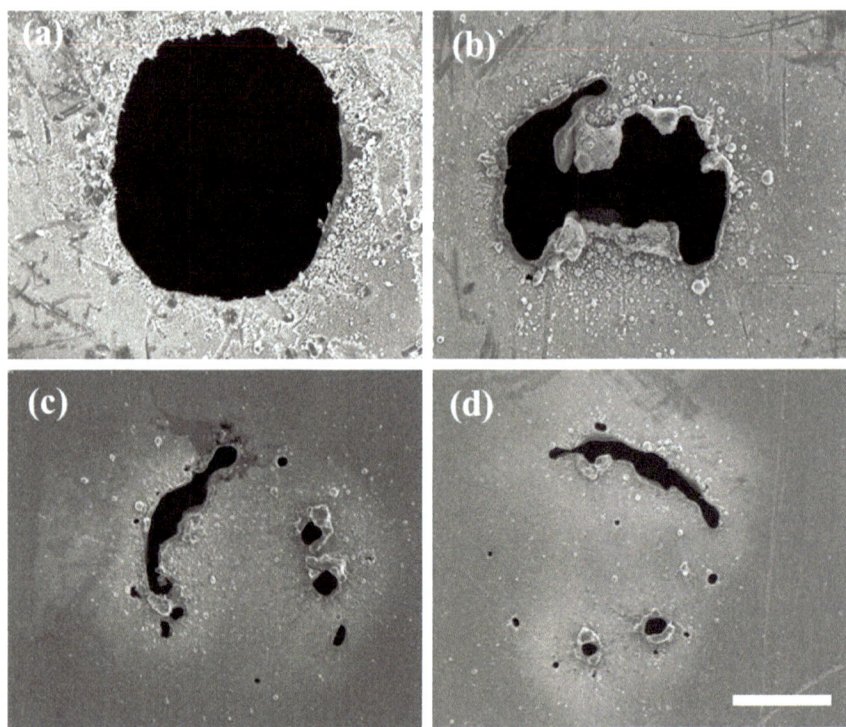

Fig. 2.15 Effect of the incident laser scanning speed on the exit area of the pilot holes machined by the LBMM process. **a** 50 mm/s laser scanning speed, **b** 500 mm/s laser scanning speed, **c** 950 mm/s laser scanning speed, and **d** 1400 mm/s laser scanning speed. The loop count was 75, the incident laser power was 6.4 W, and the laser pulse frequency was 15 kHz for all the holes. Scale bar = 100 μm [14]

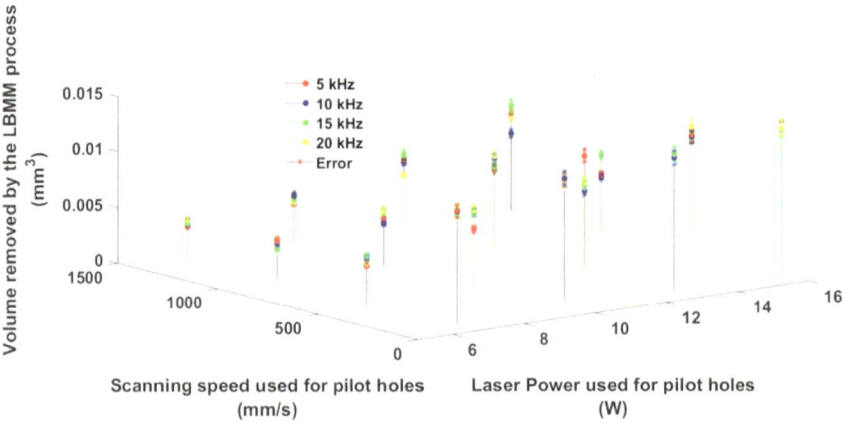

Fig. 2.16 Volume of material removed by the LBMM process as a function of average laser power and scanning speed [14]

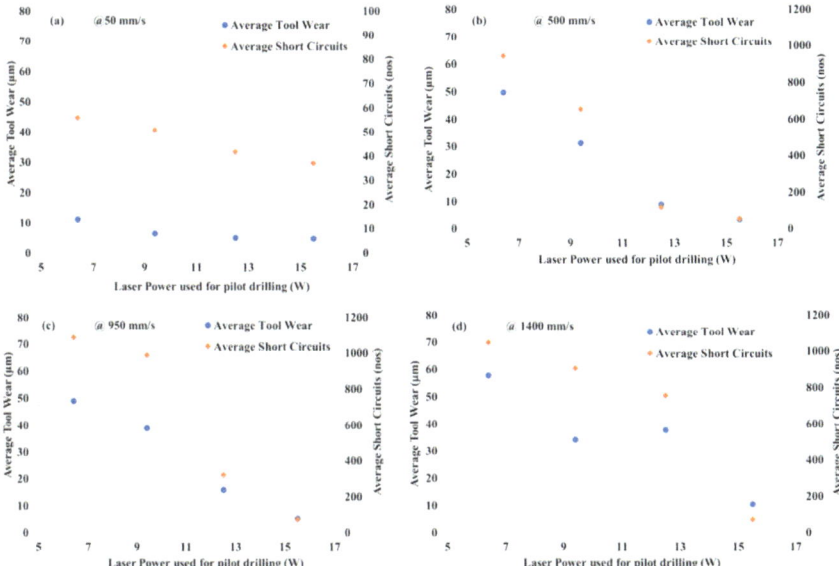

Fig. 2.17 Variation of average tool wear and number of the short circuit during the μEDM process as a function the incident laser power used for the pilot holes drilling. **a** The scanning speed was at 50 mm/s. **b** The scanning speed was at 500 mm/s. **c** The scanning speed was at 950 mm/s. **d** The scanning speed was at 1400 mm/s. The loop count was 75 for all the cases [14]

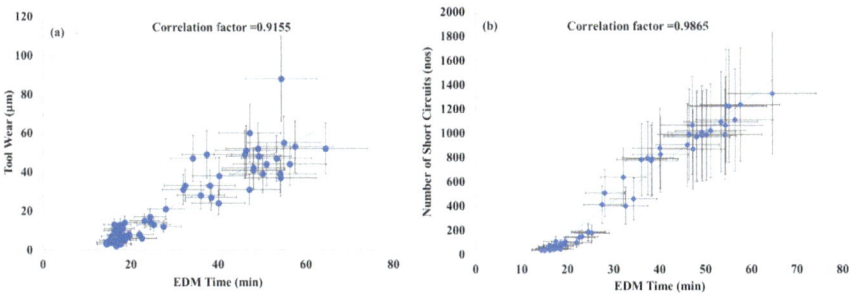

Fig. 2.18 Correlation between tool wear and short circuits vs μEDM time. **a** Tool wear vs μEDM time. **b** Short circuits vs μEDM time. Error bar shows the experimental variation [14]

2.3.1.3   Study of Residual Spatter Zone for Laser-MEDM Process

During the sequential LBMM-EDM micromachining, it is anticipated that the zone containing residual spatter caused by the LBMM process [20] will be completely eliminated. Nevertheless, our research indicates that the μEDM technique would not be capable of entirely eliminating the spatter zone if a pilot hole is drilled using LBMM at a currently programmed diameter of 200 μm while a high laser power (15.5 W) and a low scanning speed is employed. There are two potential solutions to

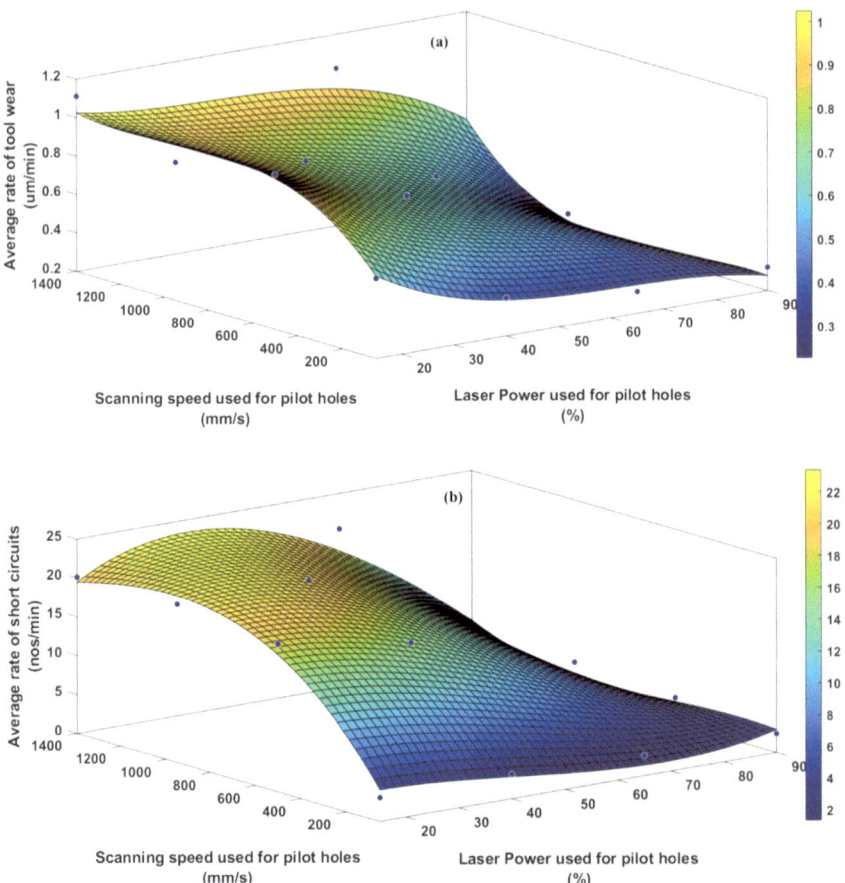

Fig. 2.19 Average tool wear rate and the short circuit occurrence rate as a function of laser incident power and laser scanning speed (used for pilot hole drilling using the LBMM process). **a** Rate of tool wear and **b** rate of short circuit occurrence [14]

this issue: as proposed by Schaeffer et al. [21], the initial approach involves wiping the sample plate with an isopropyl alcohol (IPA) wipe, followed by a thorough cleaning in an ultrasonic bath. By optimizing the programmed diameter of the LBMM process in the second approach, the residual spatter zone produced by the LBMM could be completely eliminated during the finishing operation by μEDM. We determined through experimentation that μEDM could totally eliminate the remaining spatter zone when 100 μm was utilized for the LBMM (laser power 6.4 W, scanning speed 200 mm/s, pulse frequency 20 kHz).

A comparison of the two holes machined using the μEDM and LBMM-μEDM processes is illustrated in Fig. 2.20. It can be observed from the two images (Fig. 2.20) that the areas surrounding both holes are clean of any splatter. It is obvious that upon reducing the LBMMed diameter by half, the μEDM process's performance parameter is reduced. Nevertheless, it remains considerably superior to the unmodified μEDM, as illustrated in Fig. 2.20: machining time, tool wear, and the occurrence of short circuits were all reduced by 95%, a factor of 4, and 2.5 times, respectively.

The user's preference dictates whether of the two methods (utilizing an IPA wipe and ultrasonic bath or lowering the LBMMed hole size) to eliminate any residual spatter. By employing an ultrasonic bath-based cleaning technique in conjunction with an IPA wipe, customers can select more extreme settings for the LBMM procedure, such as a greater programmed diameter and high laser power. Nevertheless, in the event that consumers lack the aforementioned resources, they are encouraged to opt for a reduced diameter, diminished power, and increased scanning speed while utilizing the LBMM method.

2.3.2 Effect of Laser Parameters for Three-Dimensional Machining

Rashid et al. [22] conducted experimental study to explore the influence of different laser parameters on the overall effectiveness of the micromilling LBMM-μEDM process. Figure 2.21 displays the 3-D response bar plots produced from the experimental data for μEDM milling time. The figure shows the influence of scanning speed and laser power on the μEDM milling time. No significant trend was observed in the μEDM milling, resulting from a rise in scanning speed. Additionally, it has been observed that an increase in laser power results in a substantial reduction in the μEDM milling time, surpassing the scanning speed impact. The high-level material that comprised most of the first LBMMed channel was effectively removed due to the incident laser's strength. A more substantial LBMMed milled channel was produced as heat propagated across the material's cross-section. Conversely, this leads to a faster μEDM milling operation. Furthermore, the loop count and pulse repetition rate of an EDM milling process does not affect the duration of the machining operation [22].

2.4 Current Status of Laser-Micro-EDM-Based Hybrid Process

Many researchers and companies are currently focused on creating better hybrid laser beam micro-EDM machines to improve efficiency and precision in machining. Alongside this development work, they are also securing patents for their innovations. By combining laser technology with micro-EDM principles, they hope to advance manufacturing capabilities in various industries. This emphasis on both machine improvement and patent protection underscores the importance of innovation in shaping the future of manufacturing processes. This invention relates to the

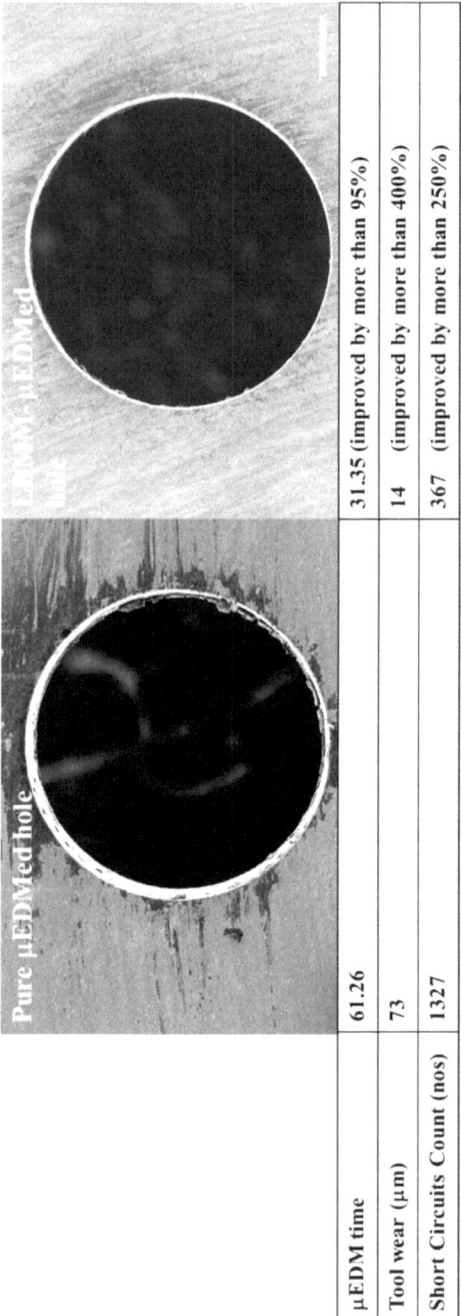

	Pure μEDMed hole	LBMM-μEDMed hole
μEDM time	61.26	31.35 (improved by more than 95%)
Tool wear (μm)	73	14 (improved by more than 400%)
Short Circuits Count (nos)	1327	367 (improved by more than 250%)

Fig. 2.20 Comparison between pure μEDMed hole and LBMM-μEDMed holes. Both holes confirm that there is no residual spatter zone present on the surrounding of the hole. Scale bar = 100 μm [14]

uEDM Milling Time vs Scanning Speed and Power

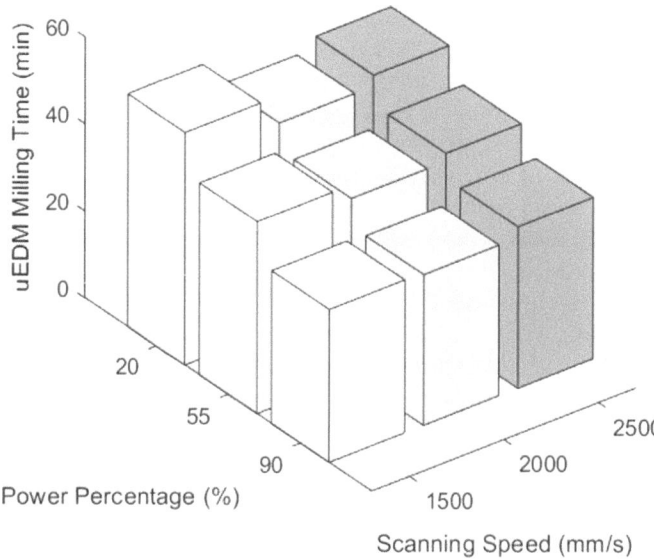

Fig. 2.21 3-D response bar plot for the μEDM milling time for variation of Power Percentage and Scanning Speed [22]

system and prototyping with fixed workpiece placement design for sequential laser drilling and μEDM machining process. The undersized holes are first formed by a laser to reduce the cycle time for penetration of the hole through the part. The part is then finished by electrical discharge machining.

U.S. Pat. No. 4,808,785 demonstrate a two-step process for laser drilling a hole in an airfoil and subsequently performing an EDM to form the diffuser-shaped part of the hole. It claims the use of a five-axis positioning system in a laser drilling workstation as a positioning reference when transferring the workpiece to the EDM workstation. The claim is also similar in the U.S. Pat. No. 6,362,446, where it claims the use of Numerical Control (NC) to control the position of the workpiece in relation to laser drill or EDM. After moving the workpiece to EDM workstation, a tooling fixture, with hole position guides is used to locate a rough blind hole.

In addition, U.S. Pat. No. 1,106,5715 also claims the use sequential machining process, which is the liquid-guided laser followed by electrical discharge machining (EDM). The workpiece is repositioned onto the EDM after the linear guided laser. All the patents have similar characteristics which the workpiece must reposition after each process.

A similar report of the invention that allows a workpiece to remain at a single location can be examined in the U.S. Pat. No. 4,857,696, issued on Aug. 15, 1986. The patent discloses the invention of having a common platform means for locating nozzle parts for programmed, sequentially accurate part transfer between a laser drilling station and a wire-EDM finishing station.

The author also proposed and patented a system and method for a two-step sequential machining process of laser drilling and microelectrical discharge machining (μEDM) on a common workpiece. A laser drill is used for drilling a rough blind hole, followed by μEDM for the finishing process. The workpiece is fixed and positioned in the μEDM workstation. A microscopic camera, which is mounted on the μEDM machine, is used to provide common coordinate position data for the laser and μEDM machine. At first, a shallow reference hole a made using the laser. Then, the microscopic camera captures and identifies the center coordinates of the reference hole. This coordinate will be used by the μEDM machine for the hole alignment for the finishing process. This invention allows the workpiece to remain at a single location for both processes.

Referring initially to Fig. 2.22, a combination laser and μEDM hole drilling station is illustrated. It includes a laser head for producing a laser output beam. Laser drilling station base is horizontally offset to the μEDM hole drilling station having a μEDM spindle head.

A μEDM hole drilling station is a three-axis CNC machine. The station has a common workpiece holder for μEDM and laser process, which is inside a tank reservoir. The station also has a microscopic camera and a μEDM spindle which are mounted on the Z-axis mounting plate.

A laser hole drilling station is a fiber laser with 50 W power. It consists of laser head and laser body. Rotary handle provides height adjustment laser head by adjusting base. A motor provides horizontal adjustment to the laser head by adjusting linear stage.

The method of setting the common coordinate position is described as follows. The workpiece is fixed into the μEDM holder. Initially, a shallow reference hole is made using laser head. Then, microscopic camera captures the hole's image and identifies the center coordinates of the reference hole. This coordinate will be used by the μEDM machine for the hole alignment for the finishing process. In conclusion, this invention allows the workpiece to remain at a single location for both processes (Figs. 2.23 and 2.24).

No.	Component	Description
1	X-axis movement motor	Responsible for X-axis movement
2	Z-axis movement motor	Responsible for Z-axis movement
3	μEDM spindle motor	Provides rotary motion for the μEDM spindle
4	μEDM spindle head	Provide a place to hold μEDM electrode
5	μEDM hole drilling station	A three-axis CNC machine
6	Tank reservoir	Holds the workpiece holder

(continued)

(continued)

No.	Component	Description
7	Common workpiece holder	Used for both μEDM and laser processes
8	Microscopic camera	Captures the image of the reference hole
9	Laser head	Produces a laser output beam
10	Y-axis movement motor	Responsible for Y-axis movement
11	X-axis gantry	Moves according to X-axis motor
12	X-axis gantry	Moves according to Y-axis motor
13	Z-axis mounting plate	Mounts the microscopic camera and the μEDM spindle
14	Laser drilling station base	Horizontally offset to the μEDM hole drilling station
15	Rotary handle	Provides height adjustment for the laser head
16	Laser body	Provides a laser source
17	Height adjustment base	Adjusts the height of the laser head
18	Laser Stage Motor	Provides horizontal adjustment for the laser head by adjusting the linear stage
19	μEDM tool	Attached to the μEDM spindle
20	Linear stage	Adjusts the horizontal position of the laser head

2.5 Summary

Chapter 2 provides a comprehensive examination of the Laser-Micro-EDM-Based Hybrid Process, offering insights into its evolution, underlying principles, challenges, and current status. Beginning with an exploration of its historical development and fundamental principles, the chapter delves into the origins of Laser-Micro-EDM technology and its integration with traditional machining techniques, marking a significant advancement in the field of manufacturing. This hybridization process, as detailed in the section on History and Principle of Laser-Micro-EDM-Based Process, has paved the way for enhanced precision and efficiency in micromachining applications.

Following the exploration of its history and principles, the chapter shifts focus to the challenges encountered during the integration of Laser-Micro-EDM technology with other machining methods. These challenges, discussed in the section on Challenges and Mitigation Techniques, include issues related to process stability, tool wear, and surface integrity. Various mitigation strategies are presented, ranging from advanced tooling designs to optimization of process parameters, aimed at overcoming these hurdles and enhancing the overall effectiveness of the hybrid process.

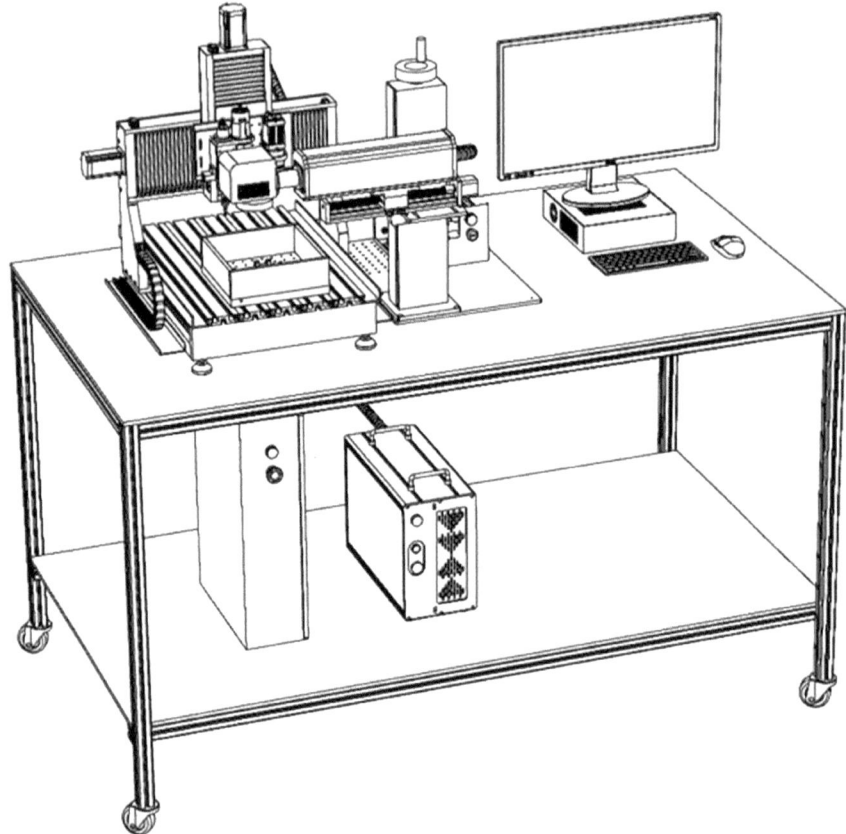

Fig. 2.22 Patented LBMM-μEDM Machine Setup

Moreover, the chapter delves into the intricate influence of laser parameters on the Laser-Micro-EDM-Based Hybrid Process. In Sect. 2.3, the effect of laser parameters on one-dimensional machining is examined, while Sect. 2.3, focuses on their impact on three-dimensional machining. Through detailed analysis and experimentation, the chapter elucidates how variations in laser parameters such as power, scanning speed, loop count and frequency can significantly affect the outcome of the hybrid machining process, influencing factors such as material removal rate and dimensional accuracy.

Finally, the chapter concludes with an overview of the current status of Laser-Micro-EDM-Based Hybrid Processes. This section provides insights into the recent advancements, ongoing research endeavors, and emerging trends in the field, highlighting the growing significance of this hybrid manufacturing approach. Overall, Chapter 2 serves as a comprehensive guide to understanding the Laser-Micro-EDM-Based Hybrid Process, offering valuable insights into its evolution, challenges, and future prospects. Part of this chapter was published in reference [14].

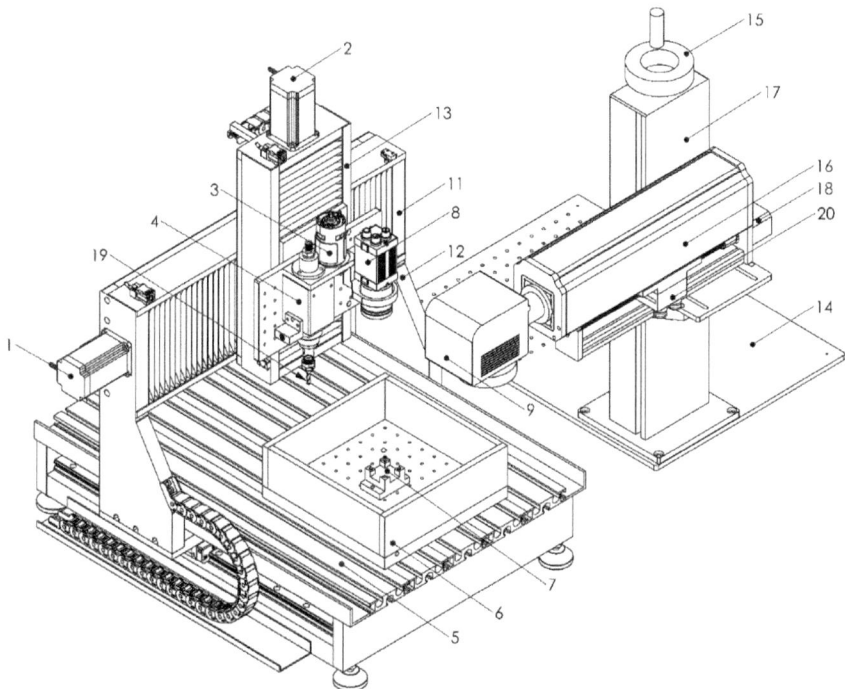

Fig. 2.23 Isometric view of a combination laser and μEDM hole drilling

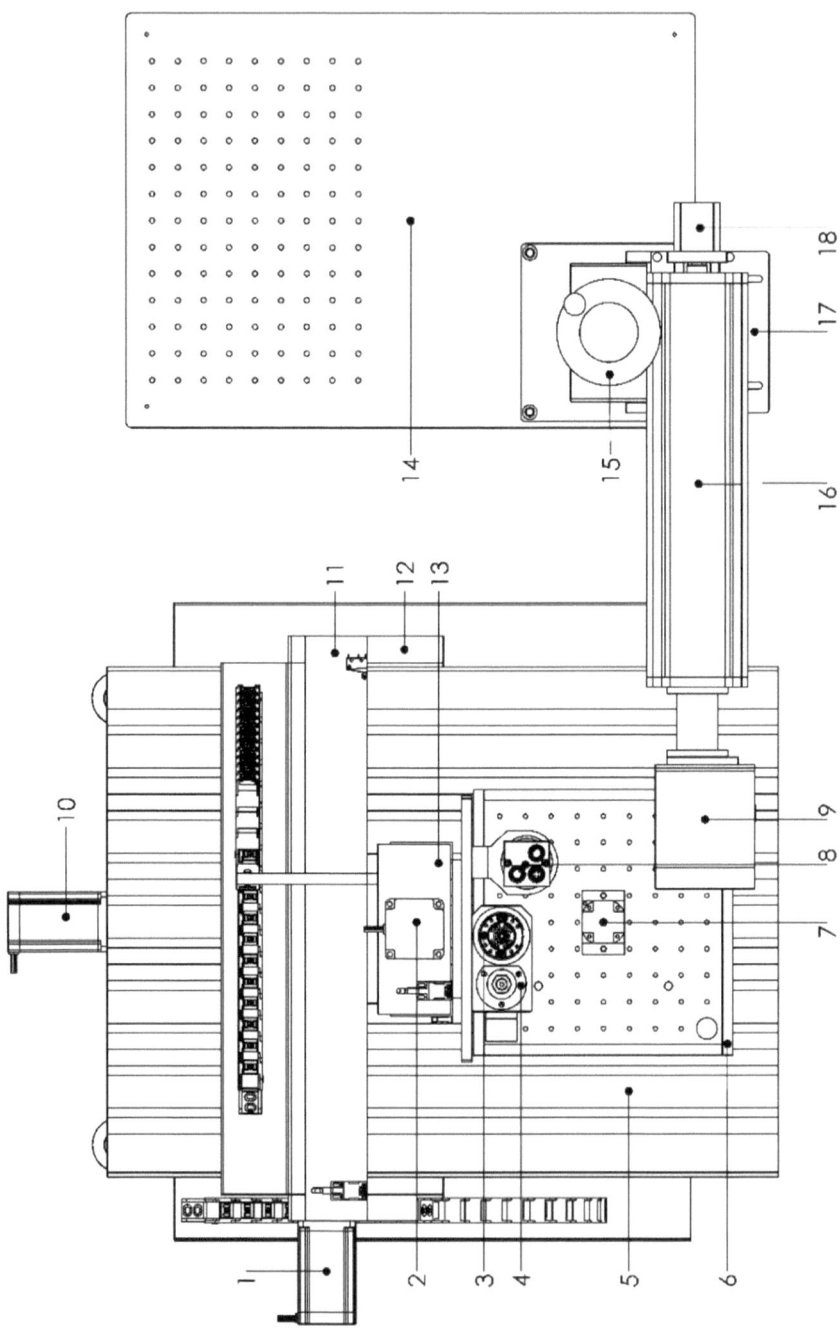

Fig. 2.24 Top view of a combination laser and μEDM hole drilling

References

1. Aspinwall DK, Dewes RC, Burrows JM, Paul MA, Davies BJ (2001) Hybrid High Speed Machining (HSM): System design and experimental results for grinding/HSM and EDM/HSM. CIRP Ann 50(1):145–148. https://doi.org/10.1016/S0007-8506(07)62091-5
2. Curtis DT, Soo SL, Aspinwall DK, Sage C (2009) Electrochemical superabrasive machining of a nickel-based aeroengine alloy using mounted grinding points. CIRP Ann 58(1):173–176. https://doi.org/10.1016/j.cirp.2009.03.074
3. Menzies I, Koshy P (2008) Assessment of abrasion-assisted material removal in wire EDM. CIRP Ann 57(1):195–198. https://doi.org/10.1016/j.cirp.2008.03.135
4. Shrivastava PK, Dubey AK (2014) Electrical discharge machining–based hybrid machining processes: a review. Proc Inst Mech Eng Part B J Eng Manuf 228(6):799–825. https://doi.org/10.1177/0954405413508939.
5. Maity KP, Choubey M (2019) A review on vibration-assisted EDM, MICRO-EDM and WEDM. Surf Rev Lett 26(05):1830008. https://doi.org/10.1142/S0218625X18300083
6. Singh R, Melkote SN (2007) Characterization of a hybrid laser-assisted mechanical micro-machining (LAMM) process for a difficult-to-machine material. Int J Mach Tools Manuf 47(7–8):1139–1150. https://doi.org/10.1016/j.ijmachtools.2006.09.004
7. Sun A, Chang Y, Liu H (2018) Metal micro-hole formation without recast layer by laser machining and electrochemical machining. Optik (Stuttg) 171:694–705. https://doi.org/10.1016/j.ijleo.2018.06.099
8. Arrizubieta JI, Klocke F, Gräfe S, Arntz K, Lamikiz A (2015) Thermal simulation of laser-assisted turning. Procedia Eng 132:639–646. https://doi.org/10.1016/j.proeng.2015.12.542
9. Ding H, Shen N, Shin YC (2012) Thermal and mechanical modeling analysis of laser-assisted micro-milling of difficult-to-machine alloys. J Mater Process Technol 212(3):601–613. https://doi.org/10.1016/j.jmatprotec.2011.07.016
10. Feng S, Huang C, Wang J, Jia Z (2019) Surface quality evaluation of single crystal 4H-SiC wafer machined by hybrid laser-waterjet: comparing with laser machining. Mater Sci Semicond Process 93:238–251. https://doi.org/10.1016/j.mssp.2018.12.037
11. Li L, Diver C, Atkinson J, Giedl-Wagner R, Helml HJ (2006) Sequential laser and EDM micro-drilling for next generation fuel injection nozzle manufacture. CIRP Ann 55(1):179–182. https://doi.org/10.1016/S0007-8506(07)60393-X
12. Kim S, Kim BH, Chung DK, Shin HS, Chu CN (2010) Hybrid micromachining using a nanosecond pulsed laser and micro EDM. J. Micromechan Microeng 20(1):015037. https://doi.org/10.1088/0960-1317/20/1/015037
13. Al-Ahmari AMA, Rasheed MS, Mohammed MK, Saleh T (2016) A hybrid machining process combining micro-EDM and laser beam machining of nickel–titanium-based shape memory alloy. Mater Manuf Process 31(4):447–455. https://doi.org/10.1080/10426914.2015.1019102
14. Rashid MAN, Saleh T, Noor WI, Ali MSM (2021) Effect of laser parameters on sequential laser beam micromachining and micro electro-discharge machining. Int J Adv Manuf Technol 114(3–4):709–723. https://doi.org/10.1007/s00170-021-06908-8
15. Yeo SH, Aligiri E, Tan PC, Zarepour H (2009) A new pulse discriminating system for micro-EDM. Mater Manuf Process 24(12):1297–1305. https://doi.org/10.1080/10426910903130164
16. Singh AK, Patowari PK, Deshpande NV (2017) Effect of tool wear on microrods fabrication using reverse μEDM. Mater Manuf Process 32(3):286–293. https://doi.org/10.1080/10426914.2016.1198015
17. ImageJ, "ImageJ." https://imagej.nih.gov/ij/index.html. Accessed 10 Sep 2019
18. Mahardika M, Tsujimoto T, Mitsui K (2008) A new approach on the determination of ease of machining by EDM processes. Int J Mach Tools Manuf 48(7–8):746–760. https://doi.org/10.1016/j.ijmachtools.2007.12.012
19. Negarestani R, Li L (2012) Laser machining of fibre-reinforced polymeric composite materials. In: Machining technology for composite materials, Elsevier, pp 288–308. https://doi.org/10.1533/9780857095145.2.288

20. Demir AG, Previtali B, Bestetti M (2010) Removal of spatter by chemical etching after micro-drilling with high productivity fiber laser. Phys Procedia 5:317–326. https://doi.org/10.1016/j.phpro.2010.08.152
21. Schaeffer KG (2008) Post-laser processing cleaning techniques. Industrial Laser Solutions. https://www.laserfocusworld.com/industrial-laser-solutions/article/14216607/post-laser-processing-cleaning-techniques. Accessed 11 Oct 2020
22. Rashid MAN, Saleh T, Hamid SA, Rashid MR (2023) Analysis and modelling of laser micro-EDM based on hybrid micro milling on stainless steel (SUS304) using box Behken design. IIUM Eng Congr Proc 1(1):14–18. https://doi.org/10.31436/iiumecp.v1i1.2996

Chapter 3
Modeling Technique and Performance of Laser Micro-EDM-Based Hybrid Processes

3.1 Introduction

This section is dedicated toward an exhaustive discussion on modeling of laser micro-EDM-based hybrid micromachining for 1-D and 3-D machining. The modeling of the 1-D hybrid micromachining process will be discussed in this section in two parts. They are: 1. Modeling of the LBMM-μEDM drilling with constant EDM parameter, 2. Modeling of the LBMM-μEDM drilling with variable EDM parameters. The difference between the two parts is that the second part takes into account the effect of micro-EDM parameters on the outcome of the hybrid micromachining process. After that, the modeling of LBMM-μEDM micromilling process for constant μEDM parameters is discussed along with the effect of process variables on the outcome.

3.2 Modeling of the LBMM-MEDM 1-D Drilling with Constant MEDM Parameter

In this research investigation, the sequential LBMM-μEDM microdrilling process was conducted with the process parameters given in Table Appendix 1. Raw data was gathered once the different LBMM and μEDM parameters had been characterized. The obtained data was cleaned and preprocessed before the final dataset was compiled. The cleaning and preprocessing of the raw data are crucial to modeling performance. The final dataset was primarily intended for two uses. Providing the data for analysis and identification of relevant process factors for the hybrid micromachining technique was one of the goals. The main goal was to produce the test and training datasets needed to assess and train the ANN model. The inputs and outputs of this supervised learning model are defined and stored inside the dataset. ANNs can be utilized as universal function approximators to capture the fundamental relationship between the input and output parameters, as was covered in the literature

© The Author(s), under exclusive license to Springer Nature Singapore Pte Ltd. 2025   59
T. Saleh et al., *Laser-MicroEDM Based Hybrid Micromachining*,
Manufacturing and Surface Engineering, https://doi.org/10.1007/978-981-97-8374-8_3

review part. Nevertheless, iterative training of the model was required to obtain the best accuracy. The best training algorithm should be used to get the ideal network topology and the best collection of hyperparameters for the ANN model. To create an effective and efficient ANN model, each of these phases is essential. This chapter provides a detailed overview of the approach used to construct the model.

3.2.1 Modeling Approach

After all the experimentation and characterization were finished, the final dataset comprising all the inputs and output parameters of the LBMM-μEDM-based sequential micromachining process was created. According to the literature review, artificial neural networks (ANNs) are extensively employed in a variety of micromachining-related research due to their effectiveness in lowering computing costs with good predictability [1, 2]. Therefore, it was decided that to forecast the LBMM-μEDM machining outputs, an artificial neural network (ANN) model should be implemented. Based on the experimental observation, it was observed that the output of the LBMM, namely the entry and exit region of the LBMMed hole, had a significant impact on the performance of the second-stage EDM finishing operation. Therefore, the entire modeling process was divided into two phases. The first phase is dedicated for creating an ANN model to forecast different LBMM outputs. After that, a second-stage model was created, using data from the prediction of the previous stage to estimate the various outputs of the final μEDM process, including tool wear, short circuit/arcing count, and machining time. Note that the purpose of this research was to examine how the LBMM process affects the LBMM-μEDM process outcome, hence characteristics like feed speed and discharge energy were held constant.

3.2.2 Correlation Study for LBMM Process Variables of the First-Stage ANN Model

The correlation analysis (Table 3.1) was carried out the selection of appropriate predictors for the LBMM-EDM process model. All potential inputs (LBMM) that were changed, as mentioned in Appendix 1, were considered as input for the first stage of the modeling process. All the LBMM outputs were strongly affected by the laser power and scanning speed.

The loop count exhibited a very strong positive connection ($r = 0.98$ and 0.93) with both the exit area and HAZ. Entry Area and Recast Layer were not significantly impacted by Loop Count since the Recast Layer formation mostly depends on the thermal energy received by the workpiece, which is independent of Loop Count, and the number of passes has no effect on the entry size after multiple passes initially. Nevertheless, there was no discernible association between any of the LBMM's

Table 3.1 Correlation coefficient between various LBMM inputs and outputs [3]

Correlation table				
	Output variables →			
Input variables ↓	Entry area	Exit area	HAZ	Recast layer
Laser power	0.83	0.92	0.82	0.71
Scanning speed	− 0.66	− 0.86	− 0.34	− 0.57
Loop count	− 0.09	0.98	0.93	0.09
Pulse repetition rate	− 0.13	0.13	− 0.14	0.12

output parameters and the pulse repetition rate. All of the Laser inputs were regarded as relevant process variables for the LBMM process modeling, that even though the pulse repetition rate had no discernible impact on the output parameters and that some of the input parameters did not exhibit sufficient correlation with some of the output parameters.

3.2.3 Correlation Study for MEDM Process Variables of the Second-Stage ANN Model

It should be noted that additional performance metrics, such as circularity, overcut taperness, and inner surface roughness, exist for μEDMed holes. The correlation analysis did show, however, that the μEDM's above other performance measures were not much impacted by the features of LBMMed holes (entry, exit, HAZ, and recast regions).

Figure 3.1 illustrates the relationship between several properties of LBMMed holes and the performance indicators (machining time, short circuit/arcing count, and tool wear) of the final finishing process, i.e., μEDM. More strongly colored matrix blocks in Fig. 3.1 indicate stronger correlations between a particular pair of variables. Red and blue, respectively, represent positive and negative associations. Figure 3.1 makes it clear that the recast layer, HAZ, and entry/exit area have a big impact on the different performance metrics of the ensuing μEDM process.

The scatter plot, as seen in Fig. 3.2, provides additional confirmation of the previously mentioned conclusions. Figure 3.2a–l makes it evident that the various quality factors of the subsequent EDM fine drilling process, such as machining time, short circuit/arcing count, and tool wear, have a significant nonlinear and negative correlation with the four major quality factors of the LBMM drilling process (entry area, exit area, recast layer, and HAZ). Therefore, in this modeling technique, all four features of the LBMMed holes were chosen as inputs for the second-stage model to predict the various outputs of the final μEDM process (entrance area, exit area, HAZ, and recast layer).

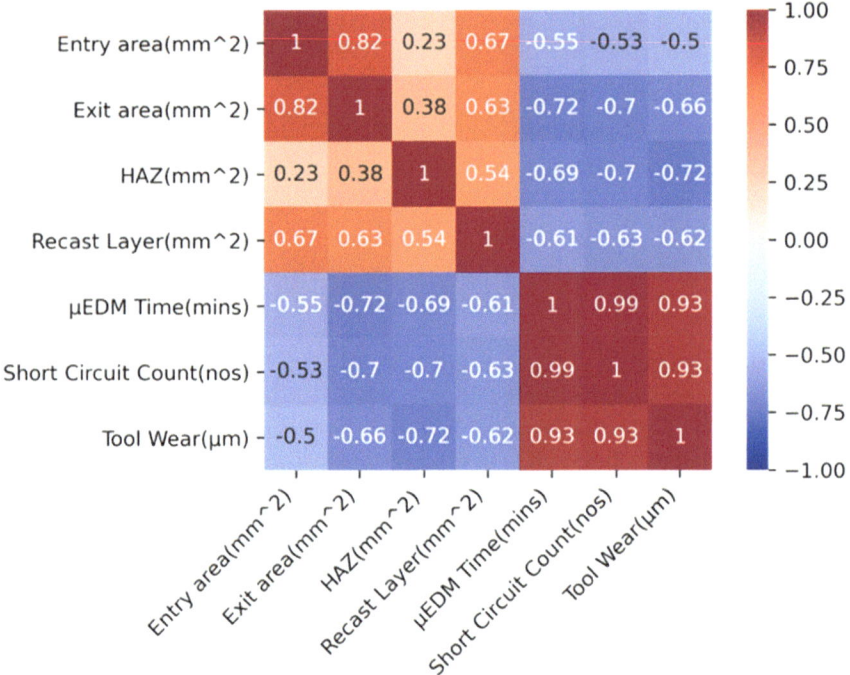

Fig. 3.1 Correlation between various LBMM output and μEDM output [3]

3.2.4 Artificial Neural Network (ANN)

A modeling technique inspired by artificial neural networks (ANNs) had been used at both stages. Because artificial neural networks (ANNs) have advantages over other regression-based models, they are widely used in numerous studies relevant to micromachining. Approximating the function for nearly all complexes, including stochastic and linear processes, is a skill that ANNs possess. An artificial neural network (ANN) is made up of layers that are stacked and connected.

The general layout of an ANN-based multi-input multi-output (MIMO) model is shown in Fig. 3.3 [4]. An activation function receives inputs from the neuron and outputs something, a summation function aggregates all of the individual outputs, and webs of weights connect the neurons in neighboring layers. The activation functions sigmoid, tanh, SoftMax, and ReLu (Rectified Linear Unit) are frequently utilized. Learning the function that controls the input–output relationship of the training dataset is a step in the ANN process. The prediction accuracy of the training model is then claimed on the test dataset. The accuracy or estimate of the model's error for a robust model is approximately the same for training and test datasets. The default performance evaluating parameters set by MATLAB R2020a are training Mean squared error (MSE), testing MSE, training Correlation coefficient (R), and testing R. These are the performance parameters used to evaluate the ANN model.

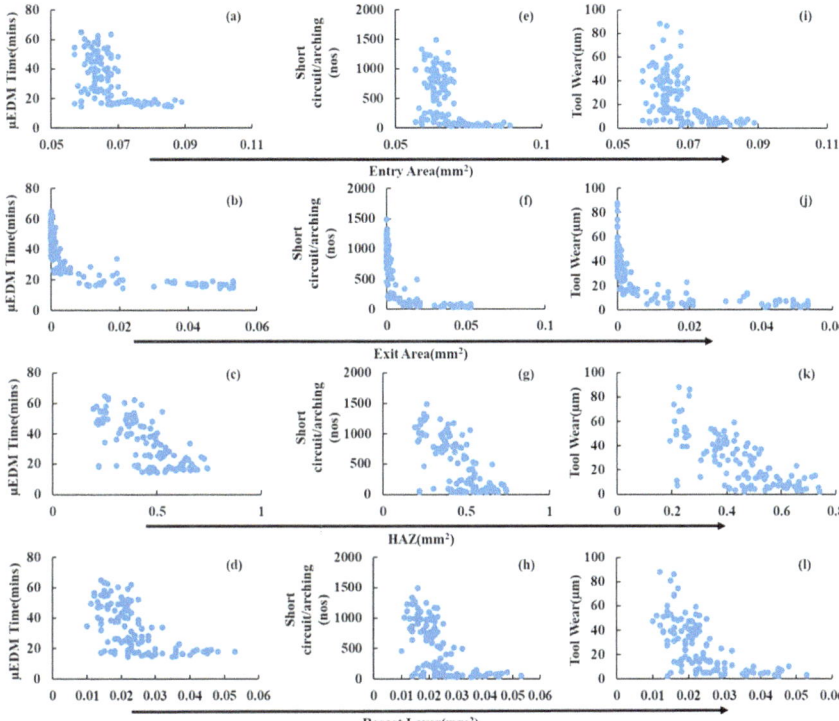

Fig. 3.2 Scatter plot to show the relationship between LBMM output and μEDM output, **a** μEDM time versus entry area of the LBMMed holes, **b** μEDM time versus exit area of the LBMMed holes, **c** μEDM time versus HAZ of the LBMMed holes, **d** μEDM time versus recast layer of the LBMMed holes, **e** short circuit/arcing count versus entry area of the LBMMed holes, **f** short circuit/ arcing count versus exit Area of the LBMMed holes, **g** short circuit/arcing count versus HAZ of the LBMMed holes, **h** short circuit/arcing count versus recast layer of the LBMMed hole, **i** tool wear versus entry area of the LBMMed holes, **j** tool wear versus exit area of the LBMMed holes, **k** tool wear versus HAZ of the LBMMed holes, and **l** tool wear versus recast layer of the LBMMed holes [3]

3.2.5 Bayesian Regularization of ANN Model

Regularization is a technique used to make adjustments to the model's various parameters in order to prevent overfitting. Compared to alternative training techniques, the Bayesian regularized ANN has the advantage of not requiring an extensive cross-validation procedure, while also maintaining robustness by removing the possibility of overfitting [5, 6]. Taking everything into account, the ANN will be trained using the Bayesian Regularization (BR) training algorithm. This is because there is a small experimental dataset—just 119 instances—that may be used to develop and test the model.

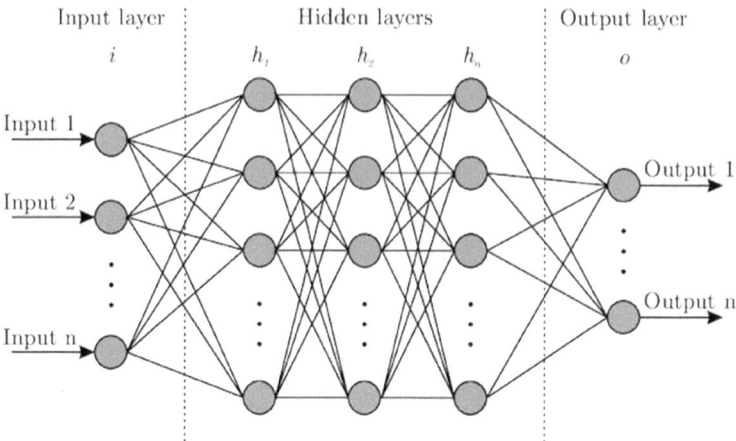

Fig. 3.3 General structure of an ANN model [4]

The Bayesian theorem of conditional probability forms the theoretical foundation of the Bayesian regularized artificial neural network. In general, for supervised learning tasks, it is possible to estimate the probability distribution function for a set of weights based on the target variable's given data. In order to approximate the most likely unknown probability distribution function of the target variable, Bayesian regularization looks for the best possible set of weights given a known probability distribution function.

Let X be the matrix containing every feature from the input. The vector y, which has N_D rows, contains the output variables, and X has N_V columns and N_D rows. Vector y can become a matrix by adding more columns. The following is a general formula for estimating a supervised learning problem's output:

$$\hat{y} = a + Xb \tag{3.1}$$

where $\hat{y}$ is the predicted value from the model. The sum of all squared error terms for the model is given by

$$E_D = \sum_{i=1}^{N_D}(y_i - \hat{y}_i)^2 \tag{3.2}$$

where the suitable set of weights can be obtained by minimizing the value of E_D.

Bayesian Regularization adds the Bayes Theorem to the error estimate for penalizing the model weights (w_j) to avoid overfitting or optimizing for local minima. The equation of the cost function $S(w)$ (analogous to E_D), expressed by hyperparameters α and β, is

$$S(w) = \beta \sum_{i=1}^{N_D} [y_i - f(X_i)]^2 + \alpha \sum_{j=1}^{N_W} w_j^2 \tag{3.3}$$

where N_w is the number of weights. The minimization of $S(w)$ results in optimized set of weights while α and β are initialized with random values.

3.2.6 Implementation and Evaluation of the Model

The suggested model, known as the "dual-stage ANN model," was a multi-input multi-output (MIMO) that utilized a combined model to predict the various target variables. As was previously indicated, the first step of this suggested model predicted LBMM outputs from a variety of LBMM inputs. In the subsequent phase, the model received inputs from various LBMM outputs to approximate μEDM machining time, occurrence of short circuits and arcing, and tool wear.

3.2.6.1 Preprocessing of Raw Data

Prior to training the sequential modeling procedure, the dataset was preprocessed using normalization based on the minimum and maximum value of each variable. The raw data transformation into normalized data was carried out as per the following equations.

$$x(i) = \frac{X(i) - \min(x)}{\max(x) - \min(x)} \tag{3.4}$$

$$y(i) = \frac{Y(i) - \min(y)}{\max(y) - \min(y)} \tag{3.5}$$

For $i = 1, 2, \ldots 119$, the $X(i)$ and $Y(i)$ represent each element of each input and output variable, respectively. And the normalized values of $X(i)$ and $Y(i)$ are $x(i)$ and $y(i)$, respectively. The minimum value of each variable is indicated by the phrases $\min(x)$ and $\min(y)$. In a similar vein, each variable's maximum values in the datasets are indicated by the words $\max(x)$ and $\max(y)$. Equations (3.4) and (3.5) state that both modeling stages will have their normalization completed. Moreover, the normalized output dataset—that is, the normalized values of tool wear and short circuit/arcing count will undergo square-root transformation in the second stage in order to eliminate any skewness in their distribution. The network may be trained more effectively and with a higher estimation accuracy thanks to the squared root transformation. The dataset will be divided into training and test datasets for each stage of the model after it is fully ready for modeling. The training set consisted of 98 randomly picked data points, with the remaining 21 being used to assess the model's performance on unseen data.

3.2.6.2   Flow of ANN Modeling

MATLAB R2020a software was utilized for the model's implementation. The architecture and full flow of the suggested dual-stage sequential model for the LBMM-EDM process are shown in Fig. 3.4. Figure 3.4 shows that the initial stage of the model (for LBMM) consists of two layers, the first of which contains three neurons and the second of which has ten. There are two layers in the second stage (μEDM), each containing two or five neurons. The output layer and the hidden layers use the linear and sigmoid functions as activation functions. The two-layer option was chosen since early training showed that adding more layers to a network did not significantly increase its prediction accuracy for both training and test datasets. Next, the model will be iterated from 1–1 topology to 20–20 topology in order to optimize the size of the hidden layers. The root mean square error (RMSE) value of each predicted variable is obtained for each combination of hidden layer size, and the average RMSE of all variables is calculated for both the training and test datasets. In the training data, a more complex topology (with many neurons) results in a lower average RMSE. However, as previously mentioned, the test dataset's average RMSE is shown to be lowest for a specific number of neurons in each hidden layer chosen for determining the network's architecture. 100 iterative trainings of the MIMO model are conducted using the optimal hidden layer size. During training, the average root mean square error (RMSE) of all the output variables (for both the training and test dataset) is calculated and saved in an array. Next, a model with a low and comparable average RMSE for both the train and test datasets is selected. The above-mentioned technique is used in both rounds of ANN modeling to choose the best-fitting model.

3.2.6.3   Estimation of Optimum Number of Hidden Layers and Neurons

The hidden layer sizes were optimized by performing up to 1000 iterations of the model, ranging from a 1–1 topology to a 20–20 topology. A total of 400 distinct neural architectures are evaluated using this approach. The root mean square error (RMSE) is computed for each projected variable and for each combination of hidden layer sizes. Additionally, the average of all the training and test RMSEs is also calculated. Subsequently, a surface plot is created to display the average root mean square error (RMSE) of the test and training sets in relation to each possible combination of hidden layer size. The optimal topology of the ANN model is indicated by the surface plot's lowest point. Since the goal is to reduce the error on the test set as much as possible, only the average RMSE of the test set is taken into account. The optimization of the most appropriate hidden layer widths for both stages is shown in Fig. 3.5a, b. The graph's $(X, Y) = (10, 3)$ point yields the lowest average RMSE value for the first stage, which is 0.006643 [6].

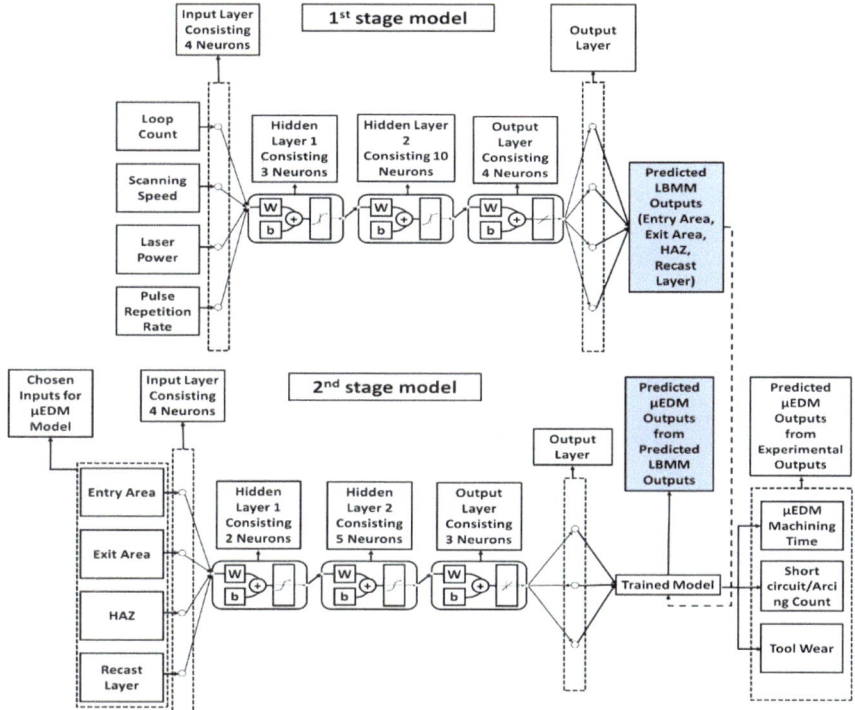

Fig. 3.4 Flow and architecture of the proposed dual-stage ANN model for the LBMM-μEDM process [3]

In contrast, the second stage is located at the graph's $(X, Y) = (5, 2)$ point and is calculated to be 0.08801. Here, the sizes of hidden layers 1 and 2 are shown by the letters X and Y, respectively. For additional modeling, the average RMSE values are considered sufficient.

Creating the best accurate model for each stage's output variables was the main goal of this research. We have created a multi-input model that forecasts several targets in two stages. To determine the most relevant process factors for the LBMM and the μEDM process, correlation analysis was first performed.

Afterward, the ANN model's inputs were selected using those variables. The output parameters of the LBMM process were regarded as inputs to the second-stage model since the μEDM input parameters were maintained constant. A graphical approach was used to determine the ideal network layout. Both models were first trained over 1000 iterations. The model with the best network design was determined to be the one that produced the lowest RMSE value after all iterations. With a 4–10–3–4 network topology, the lowest average RMSE for the first stage was measured at 0.006643. In contrast, the second stage's lowest RMSE which was evaluated at 0.08801 was discovered utilizing a 4–5–2–3 network design. By retraining the models for 100 iterations and selecting the most accurate mode, the model with the optimal

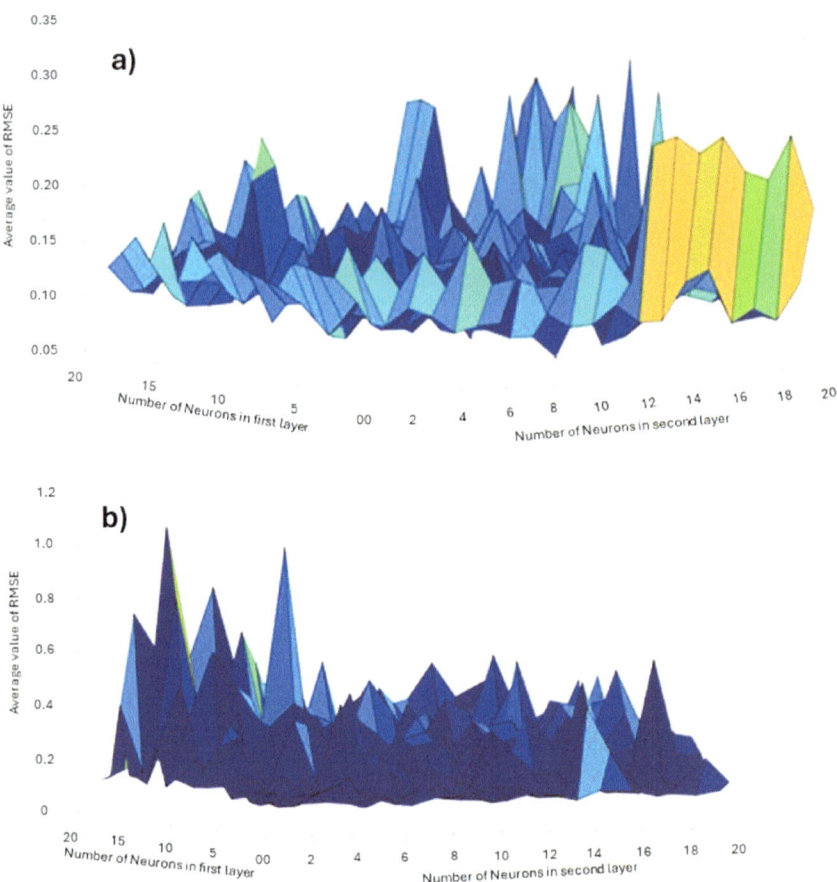

Fig. 3.5 Optimization of hidden layer size for **a** LBMM prediction and **b** μEDM prediction model based on ANN. The lowest average RMSE value for **a** was 0.006643 observed at $(X = 10, Y = 3)$ and **b** 0.08801 observed at $(X = 5, Y = 2)$ [6]

hyperparameter value was ultimately discovered. By validating the model on data that was not utilized for training, the accuracy of the model was guaranteed. The findings will be analyzed in the following part to make sure the model accurately predicts and captures the parametric effect.

3.2.7 Modeling Performance and Accuracy

Both stages of the dual-stage ANN model's modeling performance have been evaluated. Plotting the experimental response variables against the anticipated responses was done. This section addresses the performance of the different stages of the model as well as the overall performance of the model because the predicted accuracy of the first stage of the model influences the accuracy of the model overall.

3.2.7.1 Modeling Performance Parameters

The model's performance criterion used in this research is the root mean square error defined by equations below:

$$y_{iRMSE} = \sqrt{\frac{\sum_{j=1}^{N}(y_{ij_{actual}} - y_{ij_{predicted}})^2}{N}} \tag{3.6}$$

$$N_{RMSE} = \frac{\sum_{i=1}^{n} y_{iRMSE}}{n} \tag{3.7}$$

Here, y_{iRMSE} is the root mean square error for each modeled variable. $y_{ij_{actual}}$ is the experimental output of the ith variable of jth observation. On the other hand $y_{ij_{predicted}}$ is the predicted output (from the developed ANN model) of the ith variable of jth observation. N_{RMSE} is the overall network RMSE as calculated from the arithmetic average of the RMSE of all the variables of interest, and n is the number of variables. The accuracy is defined by the equations below, where $P_{accuracy_i}$ is the prediction accuracy for the ith variable and $P_{accuracy_N}$ is the overall network accuracy.

$$P_{accuracy_i} = (1 - y_{iRMSE}) \times 100\% \tag{3.8}$$

$$P_{accuracy_N} = (1 - N_{RMSE}) \times 100\% \tag{3.9}$$

Both of the model's phases have made use of the aforementioned four equations. The average laser power, scanning speed, pulse repetition rate, and loop count were the model's input variables in the first stage, and the LBMMed holes entry area, exit area, recast layer, and HAZ were the prediction variables. The entry area, exit area, recast layer, HAZ, μEDM time, tool wear, and the quantity of short circuits/arcing were the final output variables from the suggested dual-stage ANN model. In a similar vein, the input variables for the second-stage model were the outputs of the first-stage model.

3.2.7.2 Study of the Performance of the First-Stage Model

The first-stage model used the most advantageous network design to forecast the final output parameters based on the LBMM input parameters (i.e., loop count, scanning speed, laser power, and pulse repetition rate). For the first stage of the ANN model, 4–3–10–4 was the ideal topology. There were two hidden layers in this network, the first and second of which had three and ten neurons, respectively. The difference between the actual experimental values and the first-stage model's anticipated outcomes for each of the four relevant variables—entry area, exit area, HAZ, and recast layer—is displayed in Fig. 3.6. From Fig. 3.6a–d, it can be concluded that the model prediction fits the experimental data the best in the case of the exit area. For the training dataset, the individual RMSE values for the entry area, exit area, HAZ, and recast layer were 0.0763, 0.0266, 0.0649, and 0.0966, respectively; for the test set, the corresponding values were 0.0995, 0.0208, 0.0689, and 0.0721.

The prediction accuracy of the first-stage model for the training and test datasets is summarized in Fig. 3.7. Additionally, Fig. 3.7 confirms that the variable exit area ($\sim$ 97% for the exit area and $\sim$ 92% for the remaining variables) exhibits higher predictability than the other variables. The correlation coefficient between different LBMM input factors and LBMM output parameters, as shown in Appendix 1, may help to explain why the variable LBMMed holes' exit area has higher predictability. Appendix 1 shows that the exit area of the LBMMed holes has a higher correlation with most of the LBMM inputs than the characteristics of the other LBMMed holes. Nevertheless, no discernible association was observed between any of the LBMM's output parameters and the pulse repetition rate.

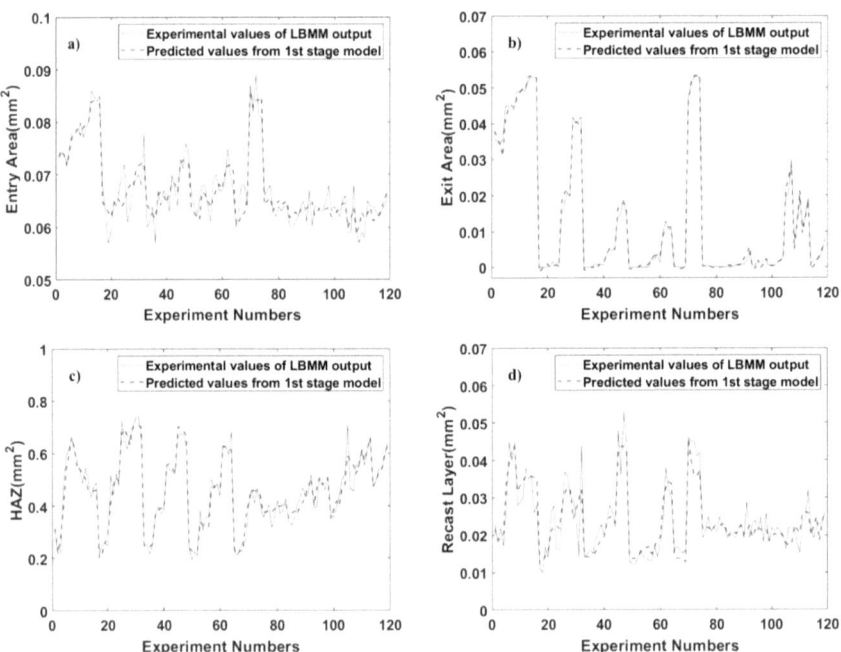

Fig. 3.6 Comparison between the actual output and predicted output for the first stage of the model, **a** entry area, **b** exit area, **c** HAZ, and (d) recast layer [3]

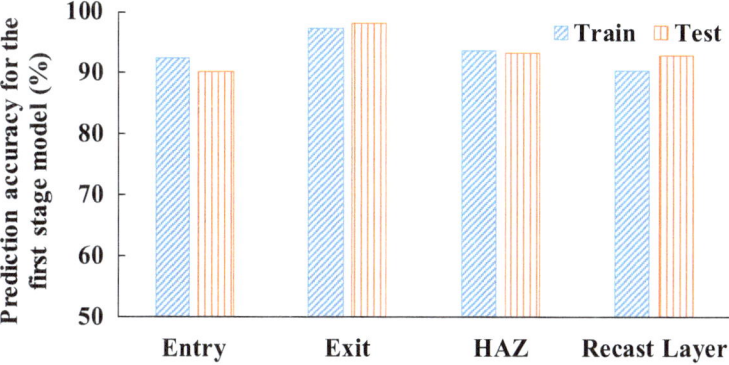

Fig. 3.7 Prediction accuracy of the four variables from the first-stage model, namely entry area, exit area, HAZ, and recast layer [3]

3.2.7.3 Study of the Performance of the Second-Stage Model

The second stage of the model acts as a link to describe how the LBMMed pilot holes' characteristics affect the output parameters of the μEDM method, thereby predicting.

The result of the sequential micromachining technique based on LBMM-μEDM. The model was trained using the experimental values of the LBMM output parameters (i.e., entry area, exit area, HAZ, and recast layer). Following the second-stage ANN, the model's ideal topology was 4–2–5–3. There were two hidden layers in this network, the first and second of which had two and five neurons, respectively. For the training dataset, the individual RMSE values for μEDM machining time, short circuit/arcing count, and tool wear were 0.1269, 0.1094, and 0.1001, respectively; for the test set, the corresponding values were 0.0932, 0.0677, and 0.0775. Figure 3.8 displays the relevant prediction accuracy for every variable. Figure 3.8 makes it clear that all of the variables' predictability falls within a similar range (~ 89 to ~ 90%). The comparison between the predicted and actual values of various variables for the second-stage model is shown in Fig. 3.9a–c, which provides additional evidence that the predictability range for all the variables in the second-stage model is similar.

3.2.7.4 Study of the Holistic Performance of the Dual-Stage ANN Model

The inputs of the LBMM (power, scanning speed, loop count, and pulse repetition rate) were utilized to forecast the outputs of the first stage (entrance area, exit area, HAZ, and recast layer) for the holistic performance. The second-stage model was then given these outputs in order to predict the toot wear, short circuit/arcing count, and EDM machining time, which are the overall performance parameters of the LBMM-EDM process. Ultimately, the estimated values of the model for the μEDM machining time, short circuit/arcing count, and tool wear were compared to the

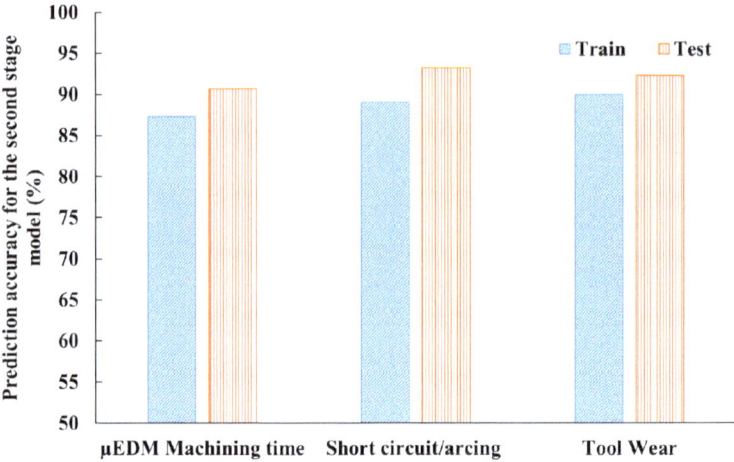

Fig. 3.8 Prediction accuracy of the three variables from the second-stage model, namely μEDM machining time, short circuit/arcing count, and tool wear [3]

actual experimental results. The root mean square error (RMSE) was computed for each variable, and the results showed that the μEDM machining time, short circuit/arcing count, and tool wear were found to be, respectively, 0.1272, 0.1085, and 0.097. Figure 3.10 displays the relevant forecast accuracy, which ranges from around 87–90%. By computing the lumped accuracy given by Eq. 3.7, Fig. 3.11 further contrasts the entire network performance (between first stage, second stage, and holistic performance).

As shown in Fig. 3.12a–c, the experimental results of the μEDM machining time, short circuit/arcing count, and toot wear were compared with the value predicted by the dual-stage ANN model. Given that the first-stage model's causality between input and output is more prominent than that of the second-stage model, it is evident that the first-stage model performs better than the latter. Furthermore, because the prediction quality of the first-stage model affects the estimation accuracy of the entire model, the dual-stage ANN model performs less holistically than the first and second stages.

3.3   Modeling of the LBMM-MEDM 1-D Drilling with Variable MEDM Parameter

The second step of this modeling was carried out similarly to the LBMM modeling—that is, EDM drilling with constant parameters—but in this case, the EDM parameters were altered. Furthermore, this model was created utilizing MIMO techniques and forecasts the different objective variables through a linked model, which is called a "dual-stage ANN model" in this study. Different LBMM outputs were predicted

Fig. 3.9 Comparison between the actual output and predicted output for the second stage of the model, **a** µEDM machining time, **b** short circuit/arcing count, and **c** tool wear [3]

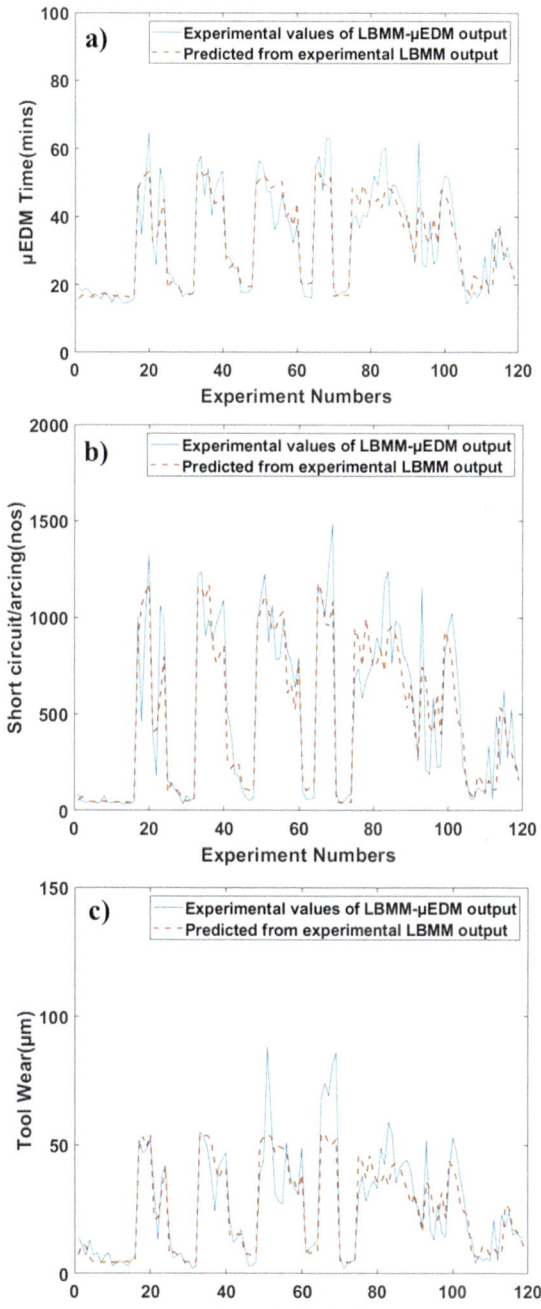

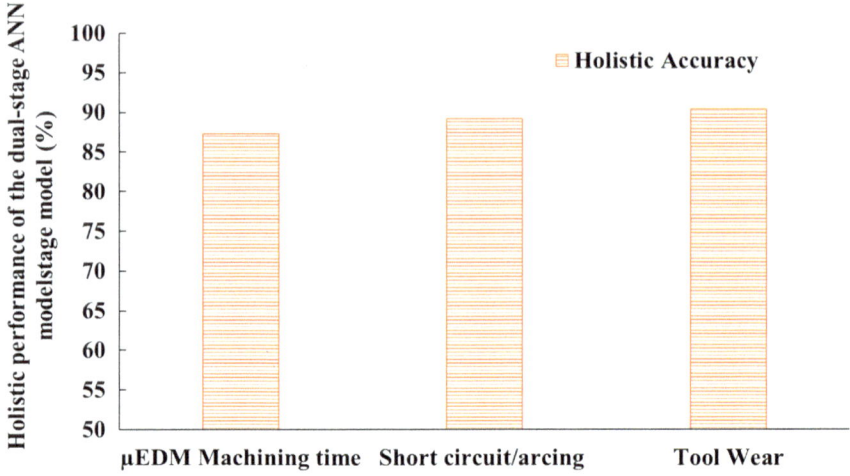

Fig. 3.10 Holistic performance of the dual-stage ANN model for μEDM machining time, short circuit/arcing count, and toot wear [3]

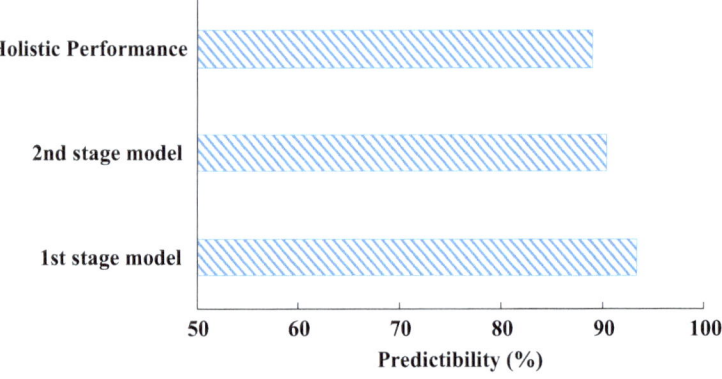

Fig. 3.11 Comparison of the performance between the first-stage model, second-stage model, and the holistic performance [3]

by the proposed model using initial LBMM inputs. The model was fed different LBMM outputs and μEDM inputs in the second step to determine how long μEDM machining will take, how likely it is that there will be short circuits and arcing, and how much tool wear there will be. Figure 3.13 guided the implementation of the model.

The correlation analysis helps determine which predictors are suitable for the LBMM-EDM process template. Throughout the first-stage model's modeling process, any potential keys in (LBMM) that were changed in line with Table 3.2 are considered input (first stage). The LBMM outputs are all significantly impacted by the laser power and scanning speed.

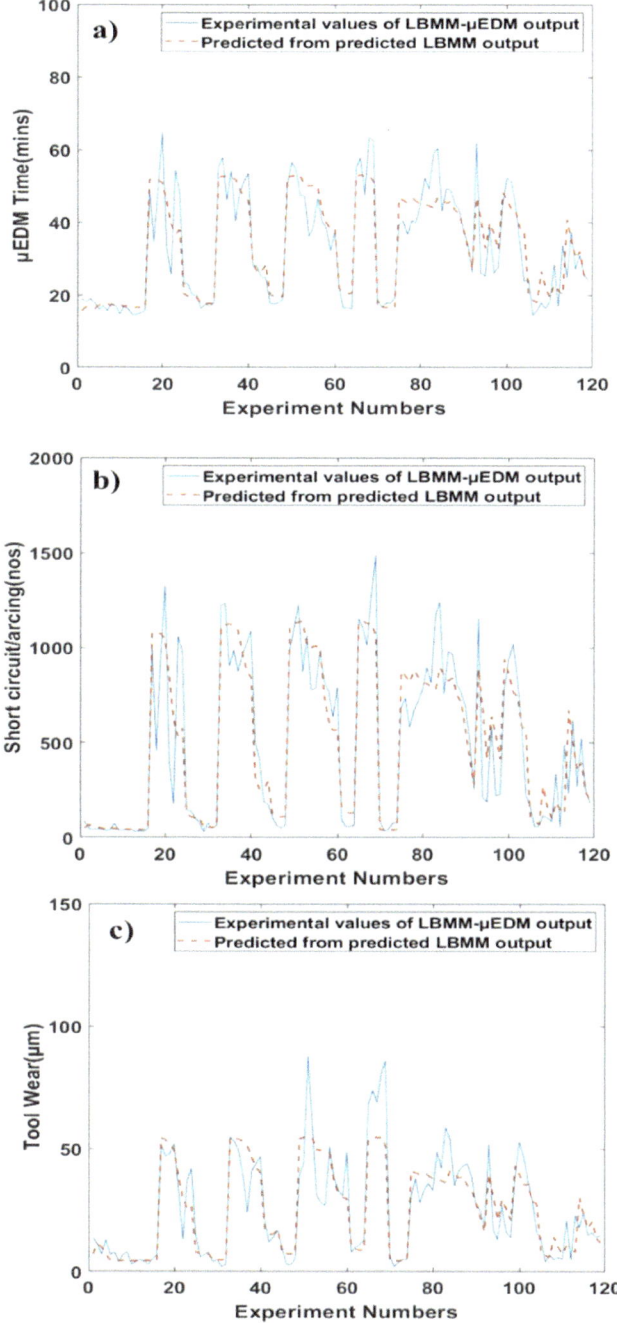

Fig. 3.12 Comparison between experimental values and estimated values by the dual-stage ANN model, **a** μEDM machining time, **b** short circuit/arcing count, and **c** tool wear [3]

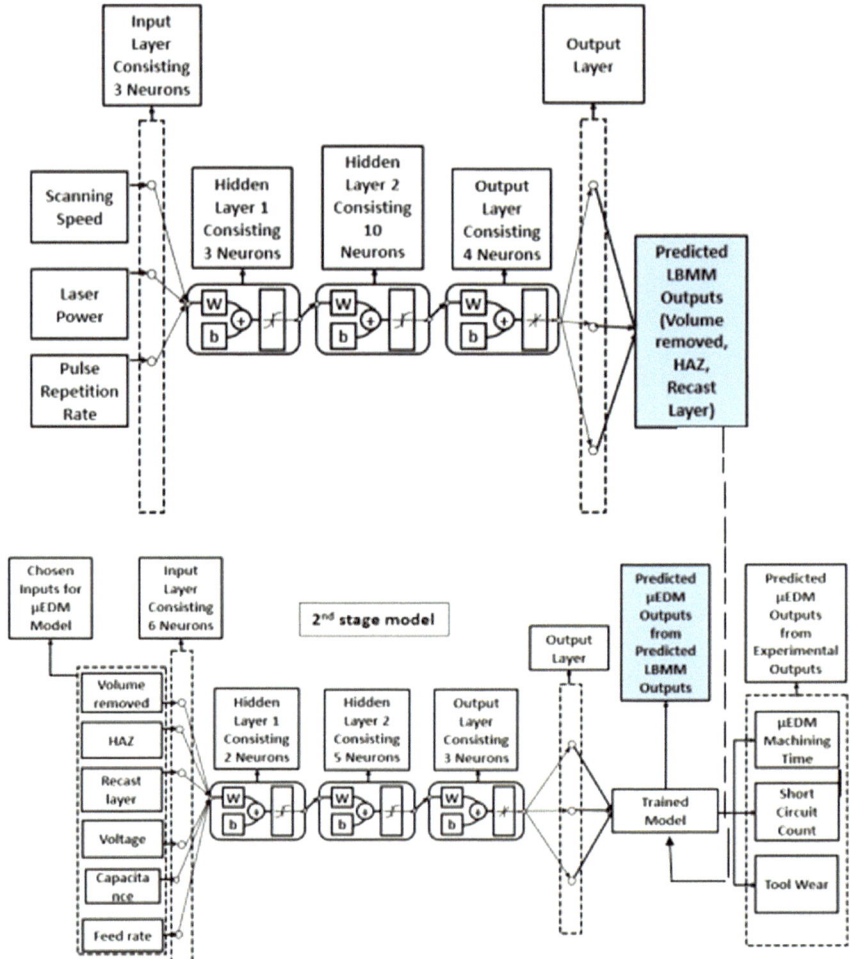

Fig. 3.13 LBMM-μEDM process dual-stage ANN model flow and architecture (drilling with Variable EDM Parameter) [7]

Table 3.2 Correlation coefficient between various LBMM inputs and outputs [7]

Correlation table

Input variables	Output variables		
	Volume removed	HAZ	Recast layer
Laser power	0.5067	0.4343	0.5139
Scanning speed	− 0.6948	− 0.3143	− 0.4951
Pulse repetition rate	0.0487	− 0.0059	− 0.0759

Three modeling approaches have been attempted in this research, according to the methodological flowchart (Fig. 3.13). Unlike the first modeling technique discussed in the previous section, an attempt was made to simulate the LBMM-μEDM microdrilling process in the second modeling, where the μEDM parameters were modified in tandem with the LBMM parameters. The overall modeling performance of the LBMM-based microdrilling, with both the LBMM and μEDM parameters modified, is described in this section.

3.3.1 Modeling Performance

In Fig. 3.13, the general modeling architecture is explained. Every step of the ANN model's modeling performance has been assessed. A scatter plot was used to compare the predicted and experimental results. This section assesses the presentation of both the individual phases of the algorithm and the gross performance, since the correctness of the model as a whole hinges on how accurate the initial step of prediction is.

3.3.1.1 Study of the First-Stage Model Performance

In the first stage of the model, the most suitable network architecture was identified using a method described in the earlier section to predict the output parameters from the LBMM input parameters (i.e., laser power, scanning speed, and pulse repetition rate). Figure 3.14 compares all three relevant output variables namely volume removed, HAZ, and recast layer—between their definite experimental values and the results predicted by the first-stage model. It is to note that in our earlier modeling technique, we used the entry and exit area of the LBMMed holes as the output, however, this time we used the removed volume as the output parameters for the LBMMed holes. The reason was it helped to converge the model with a better efficiency. Figure 3.14a–c indicates that for the case of the volume removed, the model prediction fits the experimental results the best. Individual RMSE values for the training dataset's volume removed, HAZ, and recast layer were 0.0917, 0.1291, and 0.1337, respectively, while they were 0.1105, 0.097, and 0.0914 for the test set. Additionally, Fig. 3.15 confirms that the volume removed exhibits higher predictability (91%) than the other variables. By examining Table 3.2, it is possible to determine why the variable LBMMed holes' volume removed is more predictable than other variables. As can be seen from Table 3.2 correlation, the volume removed of the LBMMed holes has a higher correlation with the majority of the LBMM inputs than the other parameters of the LBMMed holes. However, there was no noticeable correlation between any of the LBMM's output parameters and the pulse repetition rate.

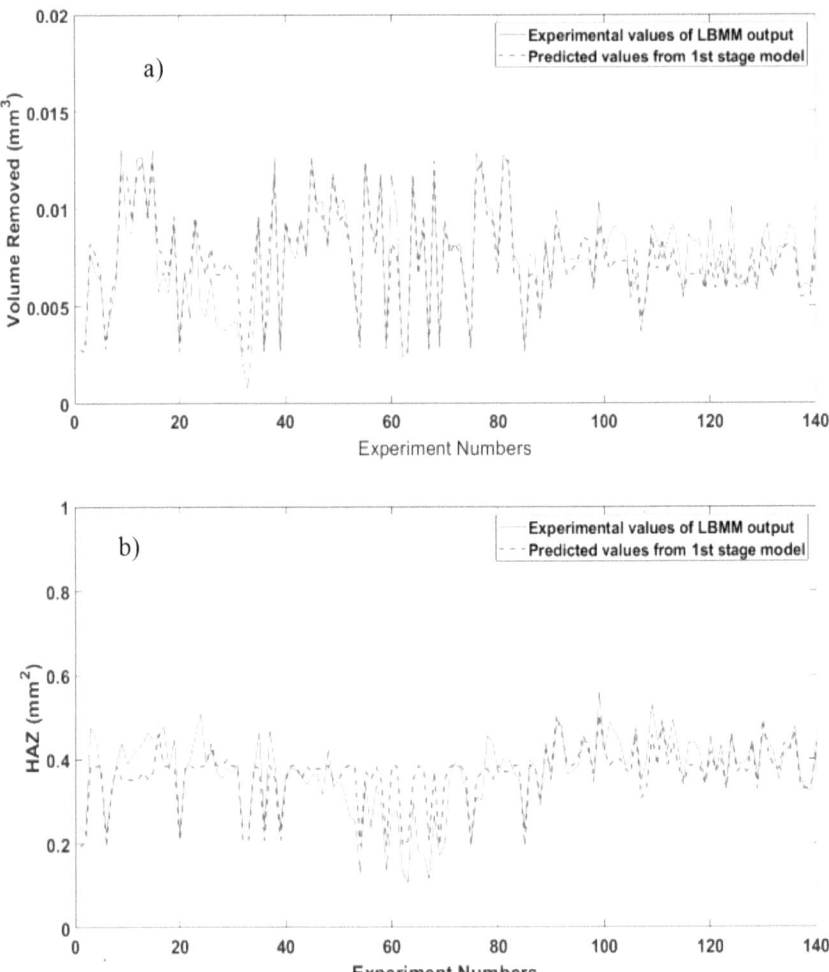

Fig. 3.14 Experimental and dual-stage ANN model comparison for the first-stage model, **a** volume removed, **b** HAZ, and **c** recast layer [7]

3.3.1.2 Study of the Second-Stage Model Performance

The second stage of the model acts as a bridge to explain how the characteristics of the LBMMed pilot holes' output parameters (removed volume, HAZ, and recast layer) as well as the various μEDM input parameters affect the output parameters of the μEDM method, which allows for the prediction of the final outcome of the LBMM-μEDM-based sequential micromachining process.

The input parameters of the μEDM process and the experimental values of the LBMM output parameters (i.e., volume eliminated, HAZ, and recast layer) were used to train the model. The training dataset's individual RMSE values for EDM

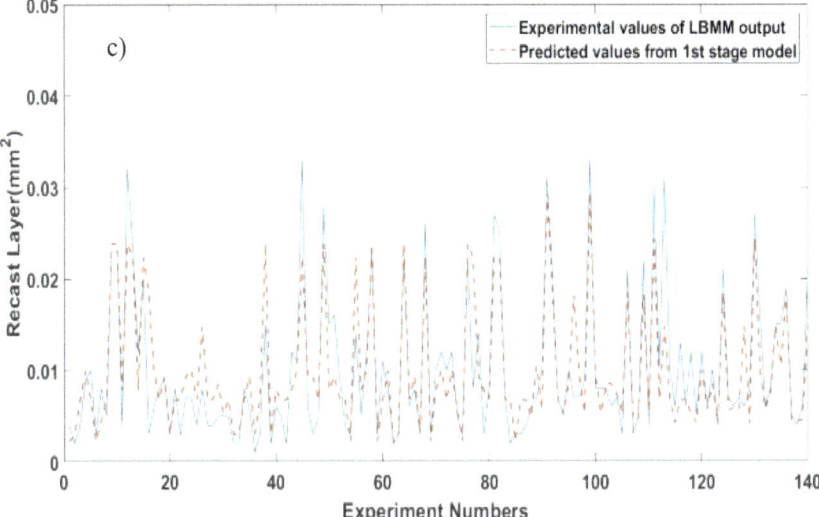

Fig. 3.14 (continued)

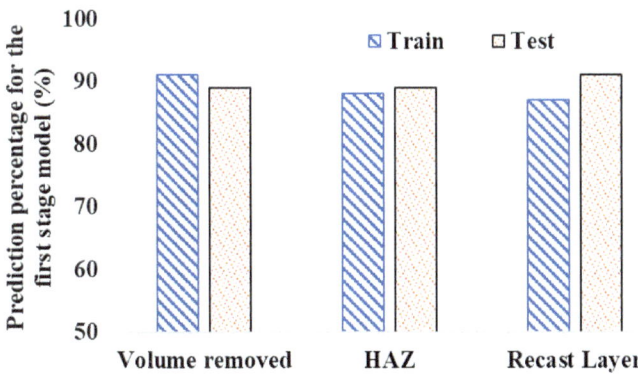

Fig. 3.15 Prediction accuracy of the first-stage model's four variables, including the volume removed, the HAZ, and the recast layer [7]

machining time, tool wear, and short circuit/arcing count were 0.944, 0.106, and 0.0929, while the test set's values were 0.0632, 0.1083, and 0.0541. The subsequent prediction accuracy for each variable is shown in Fig. 3.16. The figure shows that the predictability of each variable falls into a comparable range, which is between 90 and 94%. The definite ideals of different variables are linked to their projected values for the second-stage model in Fig. 3.17a–c, which further illustrates how reliably all of the variables fall into the same range.

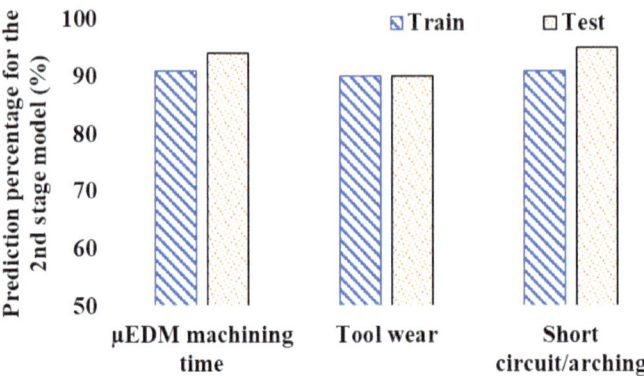

Fig. 3.16 μEDM machining time, machining instability (short circuit/arc) count and tool wear prediction accuracy from the second-stage model [7]

3.3.1.3   Study of the Overall Performance of the Dual-Stage ANN Model

To forecast the outputs of the first stage, the LBMM inputs were excluded in order to verify the proposed model's overall performance. Power, scanning speed, and pulse repetition rate (volume eliminated, HAZ, and recast layer) were among these inputs. Subsequently, the trained model was fed these projected outputs along with the μEDM input parameters to predict the performance parameters of the secondary process. These three criteria are toot wear, machining instability (short circuit/arc count), and μEDM machining time. After calculating the RMSE for each variable, it was found that the μEDM machining time, tool wear, and machining instability (short circuit/arc count) were, respectively, 0.09, 0.1271, and 0.08. Ultimately, the actual experimental results were compared with the estimated values from the model for the μEDM machining time, tool wear, and the number of short circuit counts/ arcing. Figure 3.18 illustrates the corresponding forecast accuracy, which is within the 88–92% range. An additional comparison of the entire network performance is shown in Fig. 3.19, which was obtained by using Eq. 3.8 to calculate the lumped accuracy (In between the comprehensive performance, second stage, and first stage).

3.4   Modeling of Laser-MEDM-Based Hybrid Process for 3-D Machining

In this research investigation the sequential LBMM-μEDM-based 3-D micromilling machining was conducted with the process parameter given in Appendix 3. Raw data was collected once the different LBMM-uEDM milling parameters had been characterized. The main goal of this 3-D micromilling operation was to produce a mathematical model and to test the compatibility of the experimental dataset and

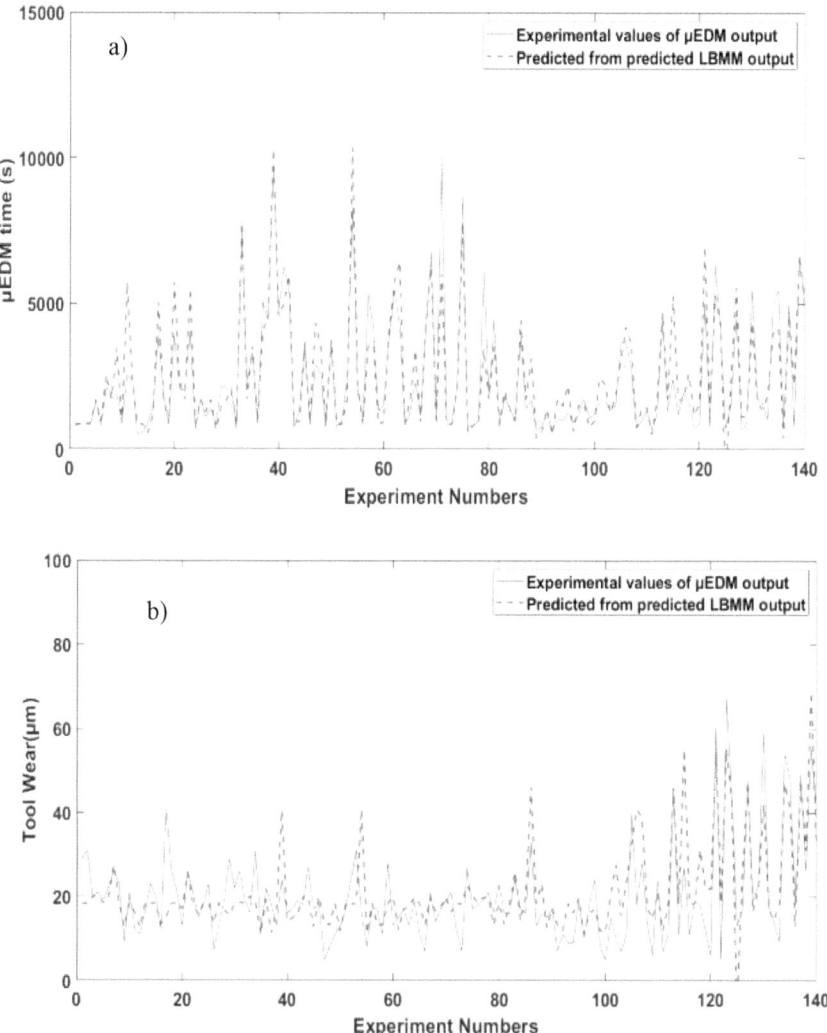

Fig. 3.17 Experimental and dual-stage ANN model comparison for second stage, **a** μEDM machining time, **b** short circuit/arcing count, and **c** tool wear [7]

predicted dataset from the Design of Experiment Tool by using the Box–Behnken designed model.

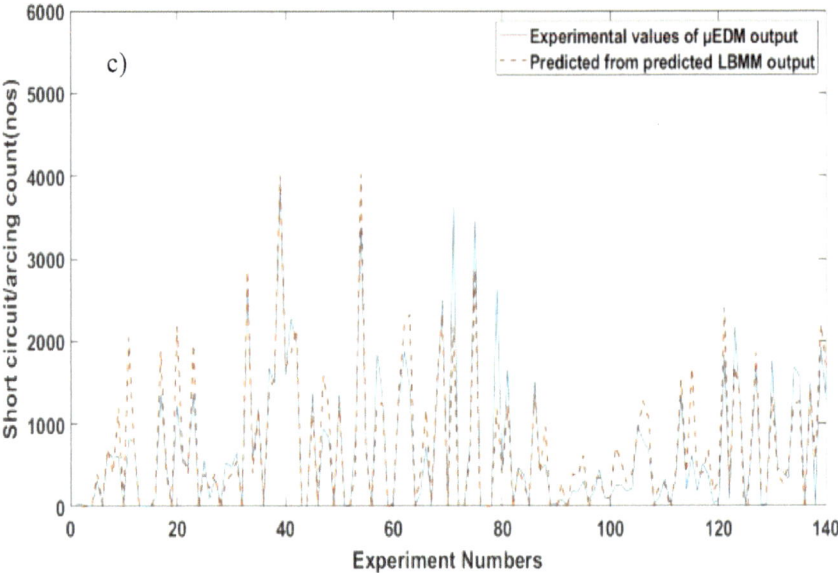

Fig. 3.17 (continued)

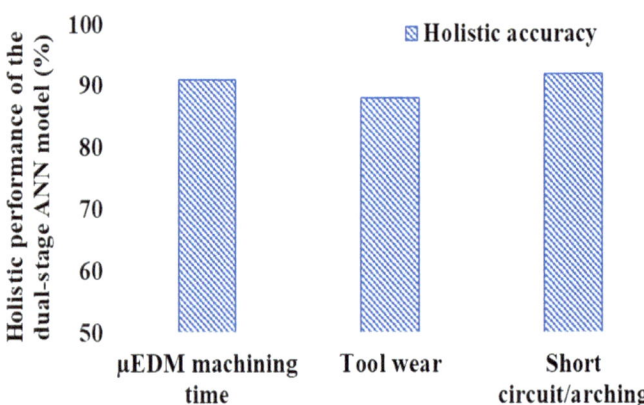

Fig. 3.18 Dual-stage ANN model's holistic performance for μEDM machining time, short circuit/arcing count, and toot wear [7]

3.4.1   Modeling of the LBMM-MEDM 3-D Milling with Constant EDM Parameter

The final phase of hybrid micromilling (LBMM-μEDM) RSM-based modeling was used with the EDM input parameters kept constant. The various laser parameters are described in Tables 3.3 and 3.4. Table 3.5 shows the complete experimental runs.

Fig. 3.19 Evaluation of the first-stage model, the second-stage model, and the overall efficiency [7]

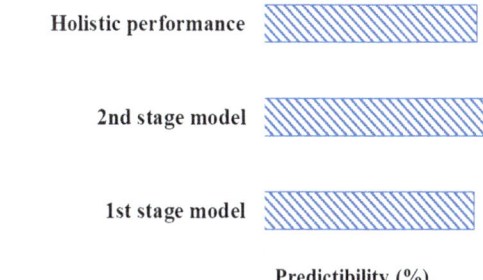

Holistic performance

2nd stage model

1st stage model

Predictibility (%)

Table 3.3 Level and parameter settings chosen for the experiments [7]

Parameter/factors	Low value	High value
Scanning speed (mm/s)	1500	2500
Power (W)	7.01	15.5
Pulse repetition rate (KHz)	5	15
Loops (nos)	5	15

Table 3.4 Values of machining variable and levels [7]

Variables	− 1	0	1
Scanning speed (mm per second)	1500	2000	2500
Power (W)	7.01	11.26	15.5
Pulse repetition rate (KHz)	5	10	15
Loops (nos)	5	10	15

3.4.1.1 Modeling of LBMM-MEDM-Based Hybrid Micromilling

In this work, the LBMM-μEDM-based micromilling technology was used to analyze mathematical modeling and optimize the processing parameters (input factors) for stainless steel (SUS 314). To ascertain the output reactions of hybrid micro-EDM machining time, tool wear, and short circuit, models were created utilizing the 3-level Box–Behnken Design (BBD) while taking into account the LBMM input elements (scanning speed, power, pulse repetition rate, and loop).

3.4.1.2 Model Development for Responses

The LBMM independent variables utilized for investigation in this study are Scanning speed, power, pulse repetition rate, and loop count. The LBMM-μEDM-based hybrid micromilling was conducted on a stainless steel workpiece (0.5 mm thickness) using the tungsten electrode (0.5 mm) in micro-EDM machining. The study examined the output responses of μEDM milling time, tool wear, and short circuits. Table 3.6

Table 3.5 Box–Behnken design of input and output responses [7]

Run	A: Scanning speed (mm/s)	B: Power (W)	C: Pulse repetition rate (KHz)	D: Loop (nos)	Response μEDM milling time (min)	Response Tw (μm)	Response short circuit (nos)
1	2000	15.5	15	10	30.96	29	47
2	2000	11.26	10	10	41.6	31	57
3	2000	15.5	10	15	34.56	26	77
4	1500	11.26	10	15	38.03	19	61
5	2000	11.26	5	5	38.93	26	77
6	2000	11.26	15	5	44.86	28	70
7	1500	11.26	10	5	41.38	29	78
8	2500	15.5	10	10	38.85	26	90
9	2500	7.01	10	10	47.02	27	68
10	2000	7.01	15	10	48.23	33	109
11	1500	7.01	10	10	45.28	31	87
12	2500	11.26	10	5	46.83	31	106
13	1500	15.5	10	10	38.36	31	74
14	2000	15.5	10	5	36.36	29	104
15	2000	11.26	5	15	38.61	23	59
16	2000	15.5	5	10	34.03	23	44
17	1500	11.26	15	10	42.9	33	79
18	2000	7.01	10	15	43.36	24	52
19	2000	7.01	5	10	52.5	41	107
20	2500	11.26	5	10	44.15	31	96
21	2500	11.26	15	10	48.71	36	85
22	1500	11.26	5	10	44.93	28	77
23	2000	11.26	15	15	36.85	26	54
24	2500	11.26	10	15	37.08	24	61
25	2000	7.01	10	5	48.68	39	97

shows numerous input and output parameters used for the modeling method both are related to the LBMM and the μEDM processes. Table 3.7 shows the total input and output responses of the LBMM-μEDM milling.

3.4.2 Response Performance of Output Models

After recording all of the experimental data into the Design Expert software, the model development was examined using the Design Expert version 13.0 software.

Table 3.6 LBMM input parameters in Box–Behnken design [7]

Run	A: Scanning speed (mm/s)	B: Power (W)	C: Pulse repetition rate (KHz)	D: Loop (nos)
1	2000	15.5	15	10
2	2000	11.26	10	10
3	2000	15.5	10	15
4	1500	11.26	10	15
5	2000	11.26	5	5
6	2000	11.26	15	5
7	1500	11.26	10	5
8	2500	15.5	10	10
9	2500	7.01	10	10
10	2000	7.01	15	10
11	1500	7.01	10	10
12	2500	11.26	10	5
13	1500	15.5	10	10
14	2000	15.5	10	5
15	2000	11.26	5	15
16	2000	15.5	5	10
17	1500	11.26	15	10
18	2000	7.01	10	15
19	2000	7.01	5	10
20	2500	11.26	5	10
21	2500	11.26	15	10
22	1500	11.26	5	10
23	2000	11.26	15	15
24	2500	11.26	10	15
25	2000	7.01	10	5

The software analyzes each response using the fit summary, model assortment, analysis of variance (ANOVA), model diagnostics, and the model graph for 3-D plots. If a model contains a lot of meaningless terms, model reduction is advised in order to strengthen the model.

3.4.2.1 Model for MEDM Milling Time

The model development for μEDM milling time follows the below steps in accordance with the Box–Behnken design procedure (Design expert version 13.0).

Table 3.7 Box–Behnken design output responses for the LBMM-μEDM milling [7]

Run	Response 1 Response μEDM milling time (min)	Response 2 Response tool wear (μm)	Response 3 Response short circuit (nos)
1	30.96	29	47
2	41.6	31	57
3	34.56	26	77
4	38.03	19	61
5	38.93	26	77
6	44.86	28	70
7	41.38	29	78
8	38.85	26	90
9	47.02	27	68
10	48.23	33	109
11	45.28	31	87
12	46.83	31	106
13	38.36	31	74
14	36.36	29	104
15	38.61	23	59
16	34.03	23	44
17	42.9	33	79
18	43.36	24	52
19	52.5	41	107
20	44.15	31	96
21	48.71	36	85
22	44.93	28	77
23	36.85	26	54
24	37.08	24	61
25	48.68	39	97

Fit Summary Statistics

The fit summary employs regression analysis to govern the causal connection between the individual variables and the chosen response. It is possible for this relationship to be linear, quadratic, or cubic. Fit summary Table 3.8 shows the μEDM milling time fit summary suggested linear model with significant terms. The highest-order linear was chosen for the μEDM machining time in order to take care of all available model terms. Therefore, the linear model with an F-value of 13.24 at $P \leq 0.0001$ was approved.

Table 3.8 Fit summary for μEDM milling time [7]

Source	Sum of squares	df	Mean square	F-value	p-value	
Mean versus total	43,518.13	1	43,518.13			
Linear versus mean	**510.85**	**4**	**127.71**	**13.24**	**< 0.0001**	**Suggested**
2FI versus Linear	39.73	6	6.62	0.6050	0.7225	
Quadratic versus 2FI	53.22	4	13.31	1.33	0.3241	
Cubic versus Quadratic	81.85	8	10.23	1.13	0.5511	Aliased
Residual	18.15	2	9.07			
Total	44,221.93	25	1768.88			

Table 3.9 Table analysis of variance for μEDM milling time [7]

Source	Sum of squares	df	Mean square	F-value	p-value	
Model	510.85	4	127.71	13.24	< 0.0001	Significant
A-SS	11.52	1	11.52	1.19	0.2874	
B-Power	431.36	1	431.36	44.71	< 0.0001	
C-Freq	0.0341	1	0.0341	0.0035	0.9532	
D-Loop	67.93	1	67.93	7.04	0.0153	
Residual	192.95	20	9.65			
Cor Total	703.80	24				

MEDM Milling Time Model Selection and Analysis of Variance

The probability (Prob > F) column for each model term in the significant models for machining time is checked to see if it falls below the 0.05 significance level, which indicates the model term is significant. Any model term that is greater than this worth denotes inconsequential terms and can be ignored from the overall model equation. Analysis of variance (ANOVA) tests were conducted to determine whether the proposed model fits the data, and the results are shown in Table 3.9. According to Table 3.9, the most significant terms were observed to be laser power and a number of the loops for modeling the secondary μEDM time.

Developed Model for the MEDM Milling Time

Equations 3.10 and 3.11 are models developed for μEDM milling time and are represented in terms of coded and actual factors. Equation (3.10), which uses coded

Table 3.10 Summary statistic for μEDM milling time [7]

Std. Dev	3.11	R^2	0.7258
Mean	41.72	Adjusted R^2	0.6710
C.V%	7.44	Predicted R^2	0.5644
		Adeq precision	12.0581

factors, predicts by comparing factor coefficients. Equation 3.11 factor equation is an analytical model used to reproduce this experiment's results.

$$Time = -20.06 + 0.9800 * A - 49.43 * B - 0.0533 * C - 2.38 * D \qquad (3.10)$$

$$Time = +58.56710 + 0.001960 * SS - 1.41239 * P$$
$$- 0.010667 * Freq - 0.475833 * Loop \qquad (3.11)$$

Adequacy of the Developed MEDM Milling Time

Validation of the developed model adequacy was performed through statistical features in order to complement the ANOVA. To verify the accuracy of the models, summary statistics on variability, lack-of-fit, R^2 adjusted and predicted R^2, adequate precision, and residual behavior were used. However, residual analysis is sufficient to determine the suitability of the model.

R-squared is a statistic that evaluates how well a model can account for outliers from the mean. The coefficient of determination (R-squared) is not always a reliable measure of model validity because it can rise when irrelevant terms are added to the model. Adj R-squared is a statistic that evaluates how well a model can account for outliers from the mean. The adjusted R-squared will drop if adding more terms to the model does not make it better. The model's predicted R-squared measures the amount of new data variation that can be ascribed to it. The metric used to assess accuracy is the signal-to-noise ratio. The average prediction error is assessed in this analysis with respect to the predicted value range at the design points.

For the model to be effective, the adjusted R-squared must be higher than 0.7 and the variation between the adjusted R-squared and predicted R-squared must not be greater than 0.2. It is preferable to have a sufficient precision ratio of at least 4, which denotes sufficient model discrimination. The different values for the μEDM milling time as produced from the design are shown in Table 3.10.

The difference is less than 0.2 because the Adjusted R^2 of 0.6710 and the Predicted R^2 of 0.5644 are reasonably in agreement. Adeq Precision measures the signal-to-noise ratio. The ideal ratio is at least 4. Here, the signal is strong enough based on the Adeq Precision ratio of 12.0581. It can be used to design the model in a certain space.

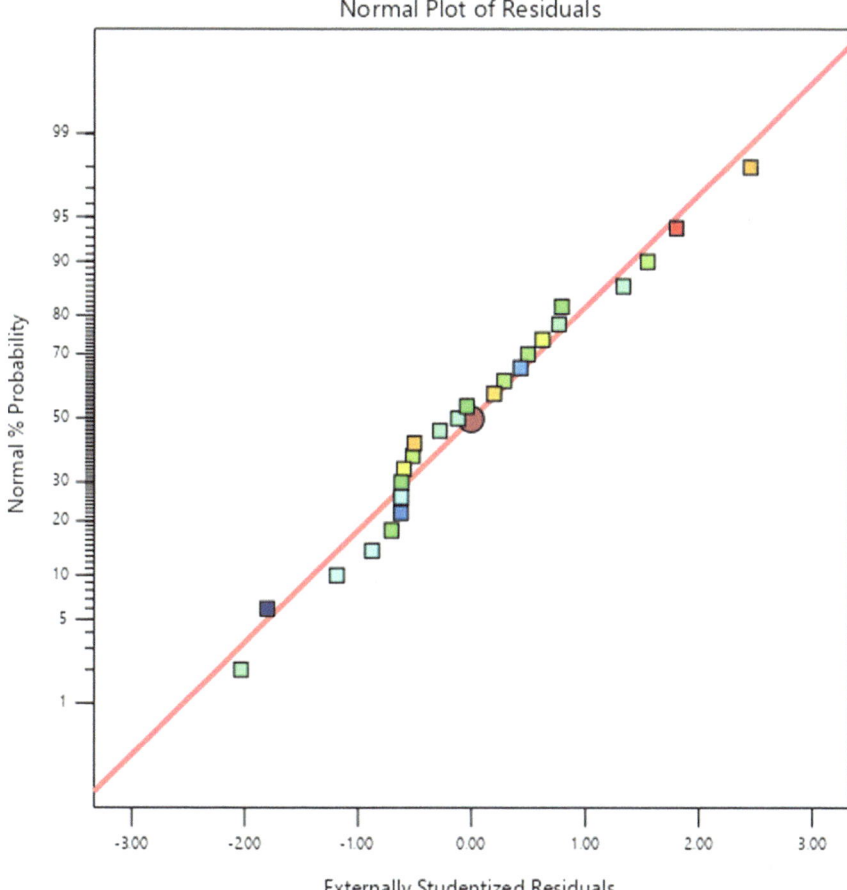

Fig. 3.20   Normal probability of residuals for μEDM milling time [7]

Examining model errors or residuals usually verifies model adequacy. Errors or residuals are the discrepancies between actual statements and regression model fitted values., i.e.

According to [8] there are several terms that residues should have for an archetypal to be suitable. These requirements are:

1.   Residues must be ordinarily scattered.
2.   The discrepancy ought to be the same.
3.   It should have zero means, and
4.   Randomized experiments.

To verify whether residuals are generally allocated, the normal plot of the residuals graph of Fig. 3.20 is plotted for the μEDM milling time.

The residual of the data obtained from the equation should be normally scattered, formless (erratically distributed), with steady variance, and zero means, according to [9]. The use of studentized residual is recommended for model adequacy tests since it takes into consideration the existence of a large residual and also large datasets that might have a high effect on the least squares fit [10]. Such analysis may also employ standardized residuals. The residuals' normal plot in Fig. 3.19 shows that they are roughly distributed randomly and follow a straight-line pattern. As a result, the μEDM milling time meets the requirement for normality.

According to [11], the normal probability plot indicates whether the residuals are distributed normally, in which case a straight line will connect the points. Figure 3.20 demonstrates that the plots have a positive value and follow a straight-line pattern. This demonstrates that the residuals are distributed normally.

It was pointed out that there are both positive and negative residuals, that a plot of residuals versus predicted values will show if the residual has the same variation or not, and for a remaining plot to be perfect and adequate, the residuals should be plotted against the predicted values. The center of the plot of residuals against what was expected must be around zero. The residual plots show that the positive and negative values are spread out randomly, and there is no clear pattern. Figures 3.21 and 3.22 show that the residual plot for constant variance checks for this experiment is acceptable, having the same variance as the original data (constant range of residuals across the graph). Figures 3.21 and 3.22 show that the residual plot for constant variance checks for this experiment is acceptable, having the same variance as the original data (constant range of residuals across the graph).

3-Dimensional Surface Plots for MEDM Milling Time

Figures 3.23 and 3.24 display the 3-D response surface plots produced by the model for μEDM milling time. The plots display the μEDM milling time trend along with changes to the important variables.

Figures 3.23 and 3.24 show a 3-D surface plot that both scanning speed and laser power influence the μEDM milling time response value. It was examined that an increase in scanning speed (ss) factors (A) increases the μEDM milling time slightly. It is also found that laser power increases and the μEDM milling time decreases, which is quite significant as compared to the scanning speed effect. The power of the incident laser was sufficient to eradicate the high-level material that was present in the majority of the initial LBMMed channel. Higher power created a larger LBMMed milled channel because heat flowed across the material cross-section. As a result, the μEDM milling process becomes faster. Figure 3.24 illustrates that the loop also has an effect on μEDM milling time. Actually, a higher loop count in the LBMMed milling process helps to eliminate additional material from the ablated zone, which enhances the μEDM milling method. Frequency/pulse repetition rate has no effect on the μEDM milling process to increase or decrease machining time.

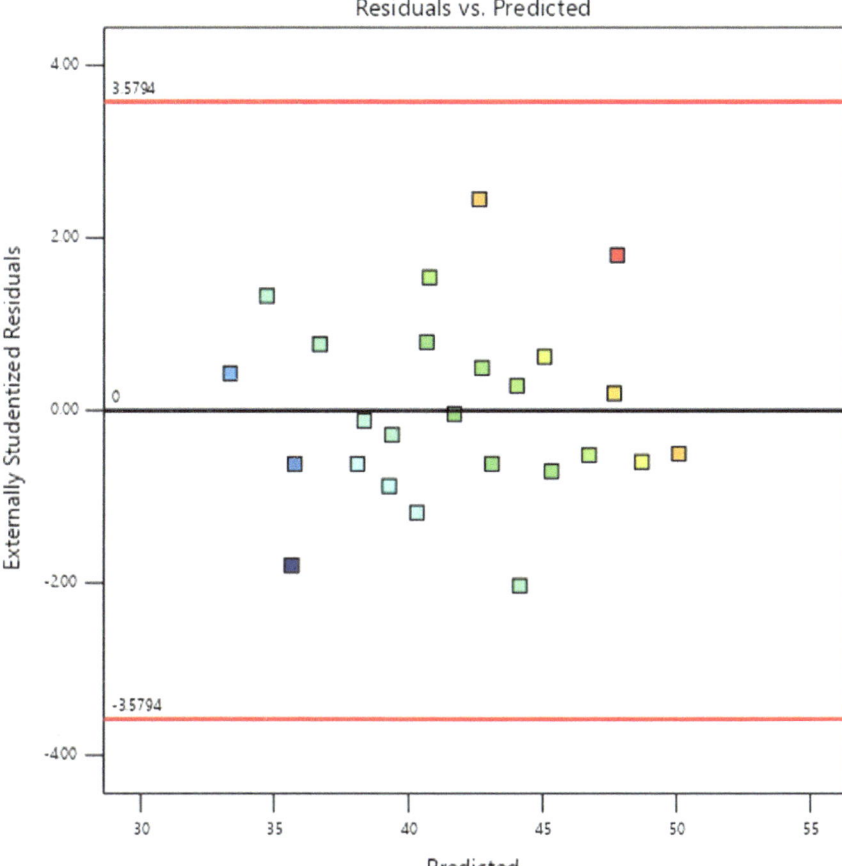

Fig. 3.21 Plot of residuals versus predicted values for μEDM milling time [7]

Study of the Model Robustness (MEDM Milling Time)

In order to estimate the robustness of the model prediction, experiments were conducted to generate more data that were not used at all for the modeling purpose. Figure 3.25 was used to show the level of fitting between the experimental and predicted μEDM time for the whole dataset. It was observed that 90% of the data for the μEDM milling time was predicted within 85% of prediction accuracy which ascertains the model robustness for the μEDM milling time.

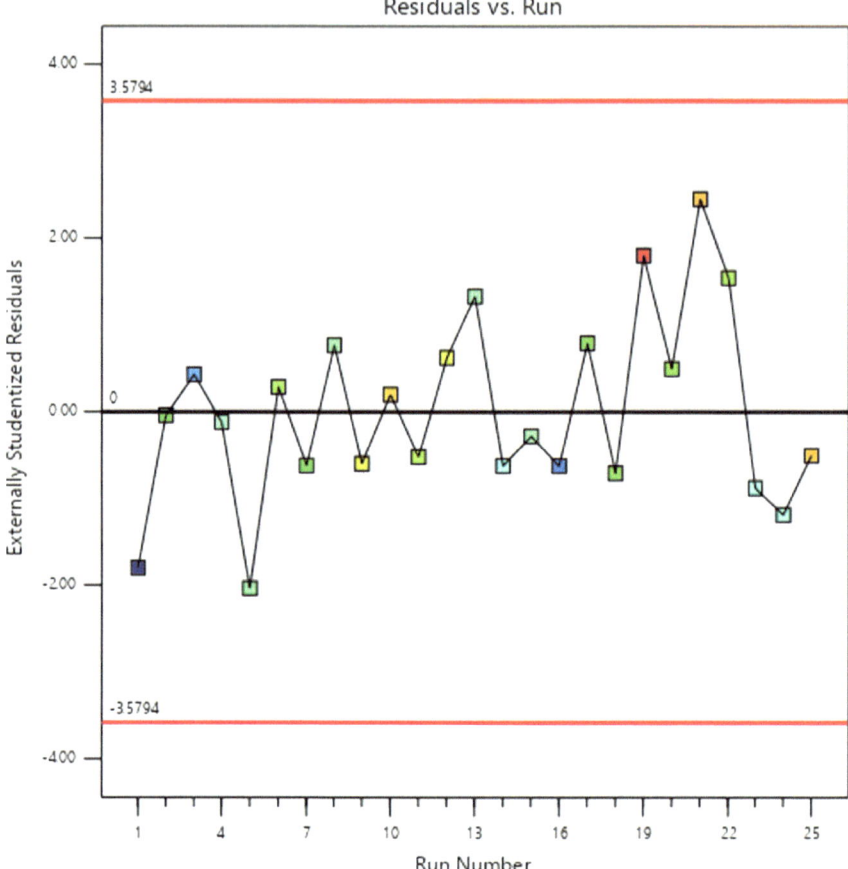

Fig. 3.22 Plot of residuals against run order for μEDM milling time [7]

3.4.2.2   Model for Tool Wear

The analysis of variance, model adequacy test, model graphs for surface plots, and model prediction competency behaviors for tool wear (μm) in LBMM-μEDM milling are presented in the following sections.

Fit Summary Statistics

Fit summary Table 3.11 shows the μEDM milling tool wear fit summary suggested linear model with significant terms. The highest-order linear is chosen for the μEDM milling tool wear in order to take care of all available model terms. As a result, the linear model that has an F-value of 3.01 and a P-value that is less than 0.0427 is accepted.

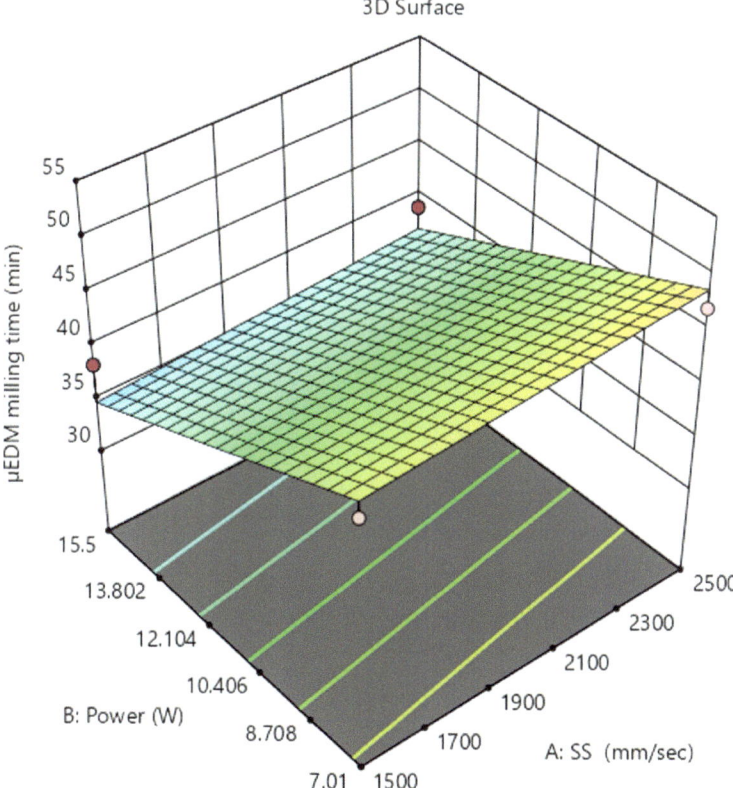

Fig. 3.23   3-D response surface plot for the μEDM milling time (power and scanning speed effect) [12]

MEDM Tool Wear Model Selection and Analysis of Variance

Analysis of variance (ANOVA) tests are displayed in Table 3.12. According to the analysis of variance (ANOVA) for μEDM milling tool wear, the model incorporating all terms is significant.

Developed Model for the MEDM Milling Tool Wear.

Equations 3.12 and 3.13 are models developed for micro-μEDM milling tool wear and are represented in terms of coded and actual factors. By determining the relative impact of the factors, typically by relating the factor coefficients, the equation in terms of coded aspects Eq. 3.12 is exploited to make projections. A projecting model is used to re-establish the experiment's results, and it is the factor equation presented in Eq. 3.13.

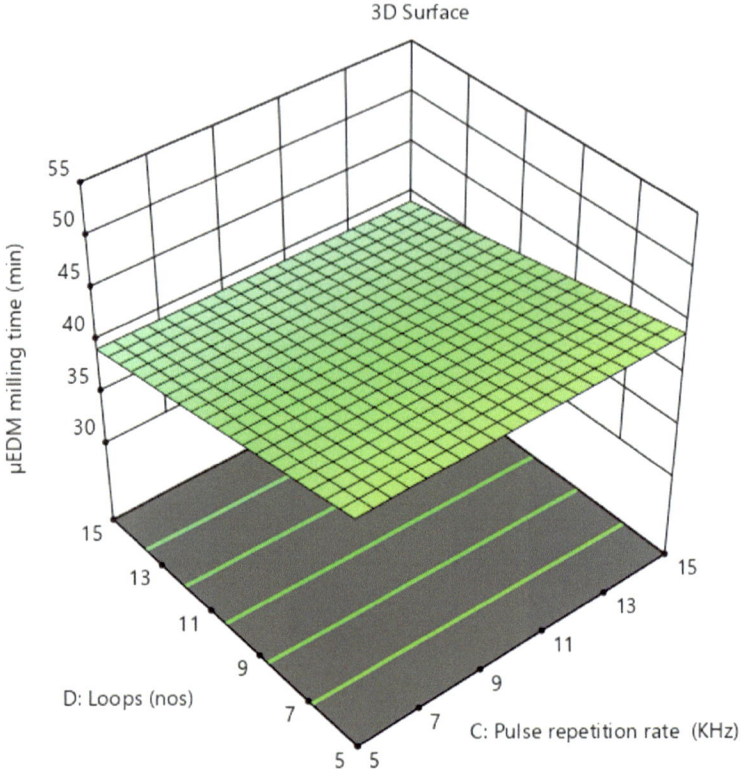

Fig. 3.24 3-D response surface plot for the μEDM milling time (loop and frequency effect) [12]

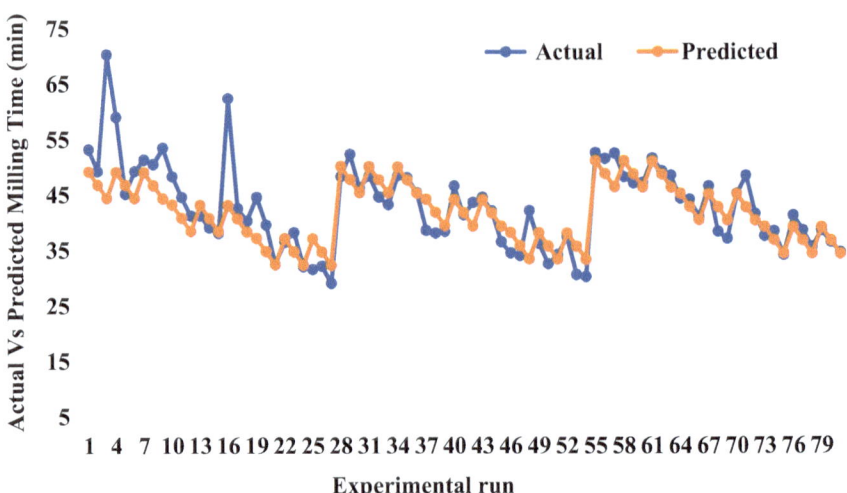

Fig. 3.25 Model competency (actual vs. predicted) for the μEDM milling time [7]

Table 3.11 Fit summary for μEDM tool wear [7]

Source	Sum of squares	df	Mean square	F-value	p-value	
Mean versus total	20,967.04	1	20,967.04			
Linear versus mean	**228.90**	**4**	**57.23**	**3.01**	**0.0427**	**Suggested**
2FI versus linear	87.83	6	14.64	0.7013	0.6535	
Quadratic versus 2FI	100.89	4	25.22	1.32	0.3281	
Cubic versus Quadratic	141.61	8	17.70	0.7119	0.7000	Aliased
Residual	49.73	2	24.87			
Total	21,576.00	25	863.04			

Table 3.12 Analysis of variance for μEDM milling tool wear [7]

Source	Sum of squares	df	Mean square	F-value	p-value	
Model	228.90	4	57.23	3.01	0.0427	*Significant*
A-SS	1.33	1	1.33	0.0702	0.7938	
B-Power	80.15	1	80.15	4.22	0.0533	
C-Freq	14.08	1	14.08	0.7411	0.3995	
D-Loop	133.33	1	133.33	7.02	0.0154	
Residual	380.06	20	19.00			
Cor Total	608.96	24				

$$\text{Tool wear} = +2.33 + 0.3333 * SS - 21.31 * \text{Power} + 1.08 * \text{Freq} - 3.33 * \text{Loop} \tag{3.12}$$

$$\text{Tool wear} = +38.98056 + 0.000667 * SS - 0.608824 * \text{Power} + 0.216667 * \text{Freq} - 0.666667 * \text{Loops} \tag{3.13}$$

Adequacy of the Developed MEDM Milling Tool Wear

Table 3.13 shows that the improved model term for μEDM milling tool wear has an R^2 of 0.3759, and the adjusted R^2 of 0.2511 and Predicted R^2 of 0.0117 are close, so the difference is less than 0.2. The adjusted R^2 for the model is significant which demonstrates the limitation of the model. However, the Adeq Precision ratio of 6.0711 indicates a strong signal.

Table 3.13 Summary statistic for μEDM milling tool wear [7]

Std. Dev	4.36	R^2	0.3759
Mean	28.96	Adjusted R^2	0.2511
C.V%	15.05	Predicted R^2	0.0117
		Adeq precision	6.0711

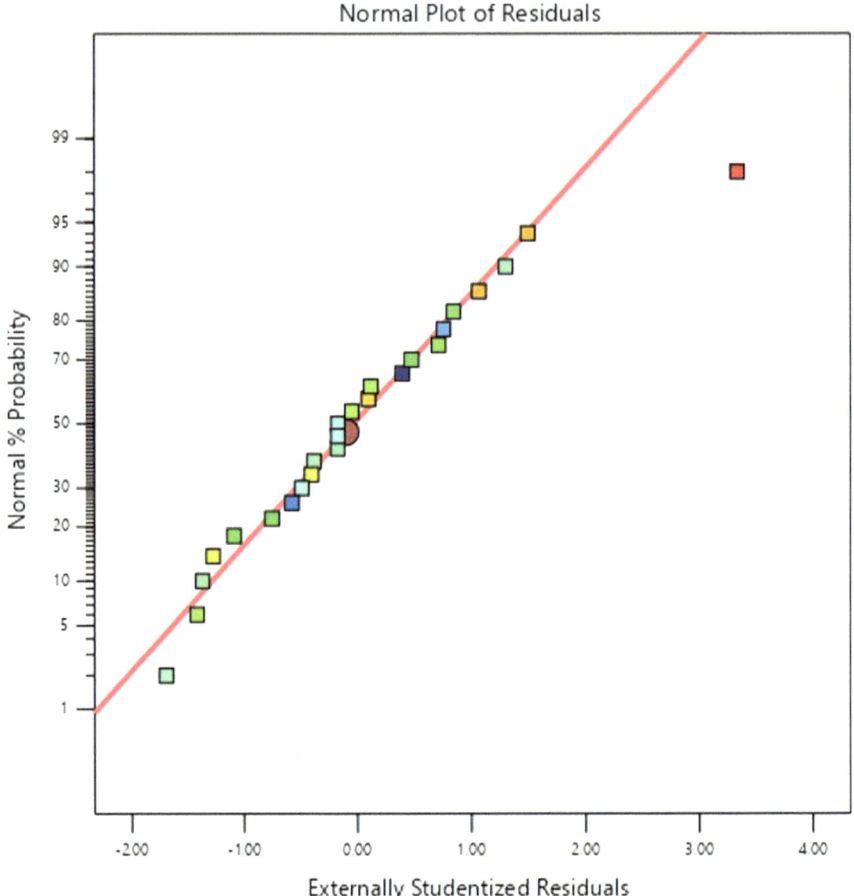

Fig. 3.26 Normal probability of residuals for μEDM milling tool wear [7]

The residuals' normal plot in Fig. 3.26 shows that they are distributed randomly and follow a straight-line pattern. As a result, the μEDM milling tool wear meets the requirement for normality.

Residual plot versus predicted values of μEDM milling tool wear shown in Fig. 3.27 signify that the data did not take any unusual structures or patterns for

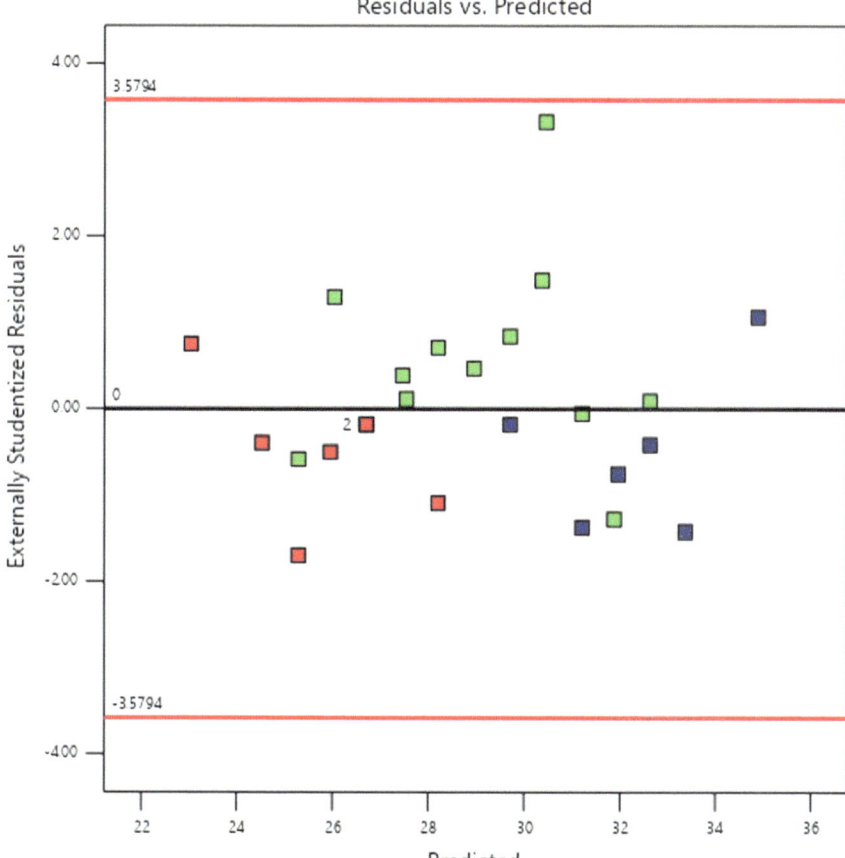

Fig. 3.27 Plot of residuals against predicted values for μEDM milling tool wear. [7]

the experimental conditions. The fitted values are randomly scattered, thus indicating that the variances of the initial observations are continuous for all standards. Hence, the constant variance hypothesis is also satisfied for the μEDM milling tool wear models.

Figure 3.28 shows the μEDM milling tool wear residual plot against run order. The plot showed no data relationship to either end. The model can predict μEDM milling tool wear because the errors are independent.

3-Dimensional Surface Plots for MEDM Milling Tool Wear

Figures 3.29 and 3.30 display the 3-D response surface plots produced by the model for μEDM milling tool wear. The plots give the μEDM milling tool wear curve with simultaneous variations in the substantial factors.

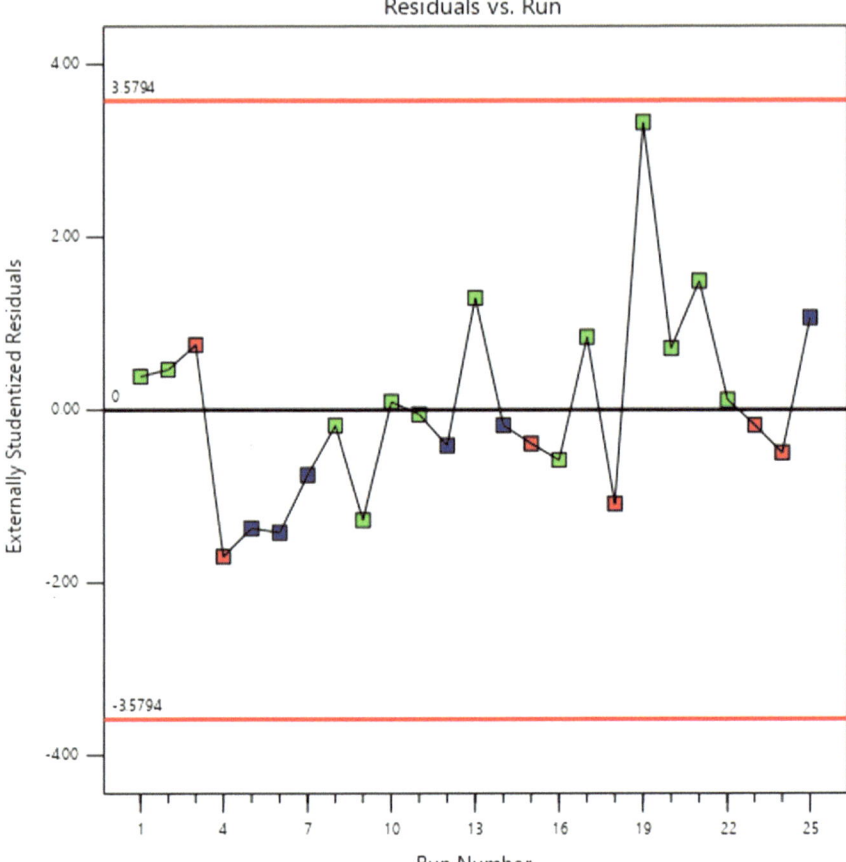

Fig. 3.28 Plot of residuals against run order for μEDM milling tool wear [7]

It is observed in Figs. 3.29 and 3.30 that laser power and loops have a significant impact on the μEDM milling tool wear. The increasing trend of power decreases the μEDM milling tool wear because the high laser power removes the workpiece material to a large extent and in the μEDM milling condition less time is required to complete the milling process. Due to this reason, tool wear during μEDM milling is decreased. Similarly, the loop number is increased means the more ablated zone is created in the LBMM milled channel which requires a shorter time to end up the μEDM milling process. It also reduces the tool wear rate in the μEDM milling.

Study of the Model Robustness (Tool Wear)

Similar to the μEDM milling time study Fig. 3.31 was used to show the level of fitting between the experimental and predicted tool wear for the whole dataset. It

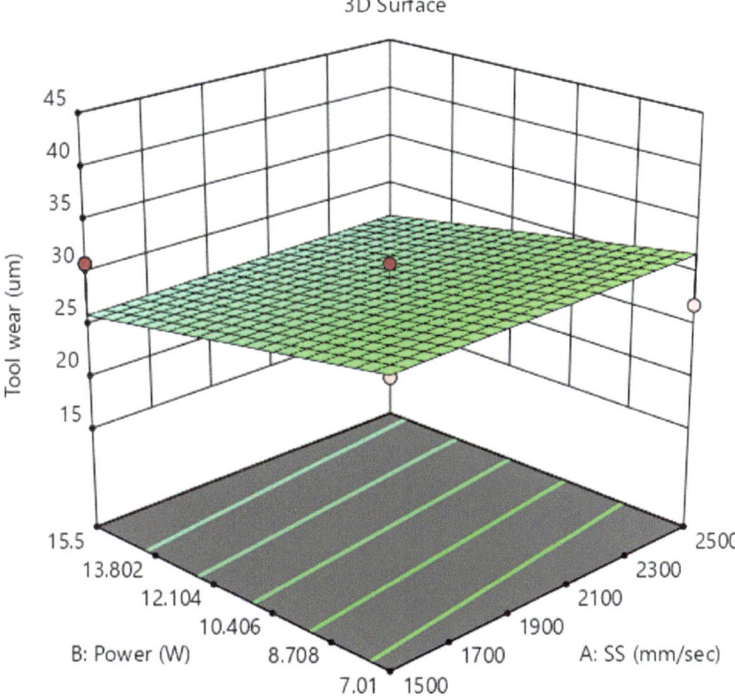

Fig. 3.29 3-D response surface plot for the μEDM milling tool wear (power and scanning speed effect) [7]

was observed that 86% of the data for the tool wear was predicted within 80% of prediction accuracy which ascertains the model robustness however not as good as for the μEDM milling time.

It was found that more than 85% of data can be explained by the predicted data.

3.4.2.3 Model for Short Circuit

In order to develop the model for μEDM milling short circuits the previous procedure was utilized. The analysis of variance, model adequacy test, model graphs for surface plots, and model prediction competency are presented for the μEDM milling short circuit model confirmation.

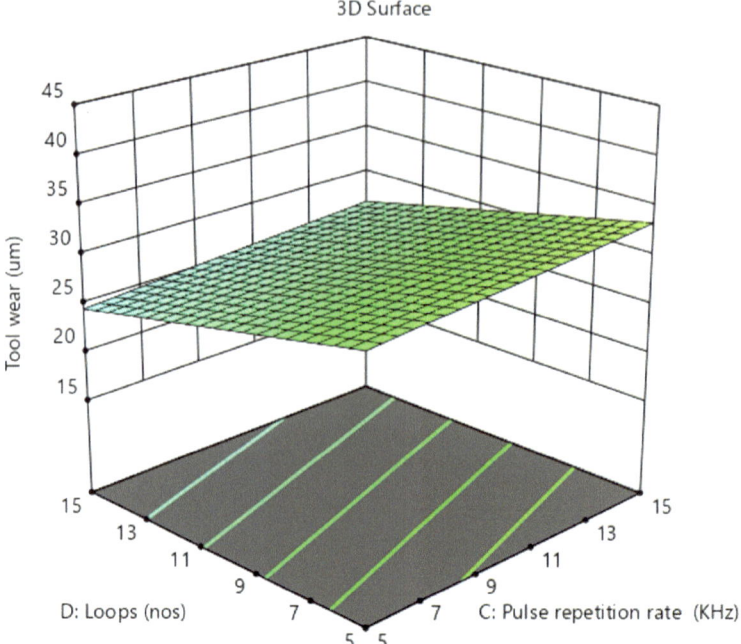

Fig. 3.30 3-D response surface plot for the μEDM milling tool wear (loop and frequency effect) [7]

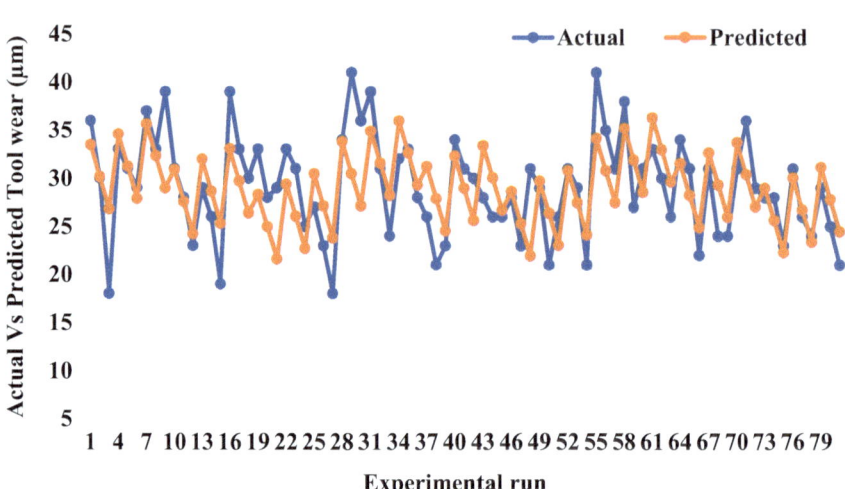

Fig. 3.31 Actual versus predicted data plot for μEDM milling tool wear [7]

Table 3.14 Fit summary for μEDM milling short circuit [7]

Source	Sum of squares	df	Mean square	F-value	p-value	
Mean versus total	1.468E+05	1	1.468E+05			
Linear versus mean	**3170.27**	**4**	**792.57**	**2.67**	**0.0621**	**Suggested**
2FI versus linear	627.32	6	104.55	0.2756	0.9392	
Quadratic versus 2FI	842.58	4	210.64	0.4715	0.7560	
Cubic versus Quadratic	3974.68	8	496.83	2.02	0.3735	Aliased
Residual	492.92	2	246.46			
Total	1.560E+05	25	6238.00			

Table 3.15 Analysis of variance for μEDM milling short circuit [7]

Source	Sum of squares	df	Mean square	F-value	p-value	
Model	3148.93	3	1049.64	3.70	0.0279	Significant
A-SS	208.33	1	208.33	0.7342	0.4012	
B-Power	588.60	1	588.60	2.07	0.1645	
D-Loop	2352.00	1	2352.00	8.29	0.0090	
Residual	5958.83	21	283.75			
Cor Total	9107.76	24				

Fit Summary Statistics

The fit summary Table 3.14 shows the μEDM milling short circuit fit summary suggested linear model with significant terms. However, the p-value is not to a lesser extent than 0.05 which indicates the model fit is poor.

MEDM Short Circuit Model Selection and Analysis of Variance

ANOVA tests are displayed in Table 3.15 Analysis of variance for μEDM milling short circuit. According to the ANOVA for μEDM milling short circuit, the model is significant by incorporating all the terms.

Developed Model for the MEDM Milling Short Circuit

The prediction of the model equation for μEDM milling short circuit profile is represented in Eqs. 3.14 and 3.15. The models are developed considering the significant

Table 3.16 Summary of model statistics for μEDM milling short circuit [7]

Std. Dev	16.84	R^2	0.3457
Mean	76.64	Adjusted R^2	0.2523
C.V%	21.98	Predicted R^2	0.0883
		Adeq precision	6.2344

terms from the improved analysis of variance in Table ANOVA. Equation 3.14 identifies the relative influence of each sort-out parameter on μEDM milling short circuit based on the coefficient of each parameter. It is observed from the calculation that the most impacting approach parameter in the order of significance is loop, power, and scanning speed. Equation 3.15 is a diagnostic model equation used to rebuild the results of the experiment.

$$\text{Short circuit} = +4.47 + 4.17 * \text{SS} - 57.74 * \text{Power} - 14.00 * \text{Loops} \quad (3.14)$$

$$\text{Short circuit} = +106.54654 + 0.008333 * \text{SS} - 1.64984 * \text{Power} \\ - 2.80000 * \text{Loops} \quad (3.15)$$

Adequacy of the Developed MEDM Milling Short Circuit

Table 3.16 is the Summary of model statistics for μEDM milling short circuit. It shows that the improved model term for μEDM milling short circuit has an R^2 of 0.3457, Adjusted R^2 of 0. 2523, and Predicted R^2 of 0.0883 are close, so the difference is less than 0.2. The adjusted R^2 for the model is significant which demonstrates the limitation of the model. However, the Adeq Precision ratio of 6.2344 indicates a strong signal.

Figure 3.32 displays the common plots of the residuals, showing that the data residuals largely adhere to linear patterns.

The residual plot versus predicted values of μEDM milling short circuit shown in Fig. 3.33 signifies that the data did not take any unusual structures or patterns for the experimental conditions. The fitted values are randomly scattered, thus indicating that the variances of the earliest observations are continuous for all values. Hence, the constant variance hypothesis is also fulfilled for the μEDM milling short circuit models.

Figure 3.34 shows μEDM milling short circuit residual plot against run order. The plot showed no data relationship to either end. The model can predict μEDM milling short circuits because the errors are independent.

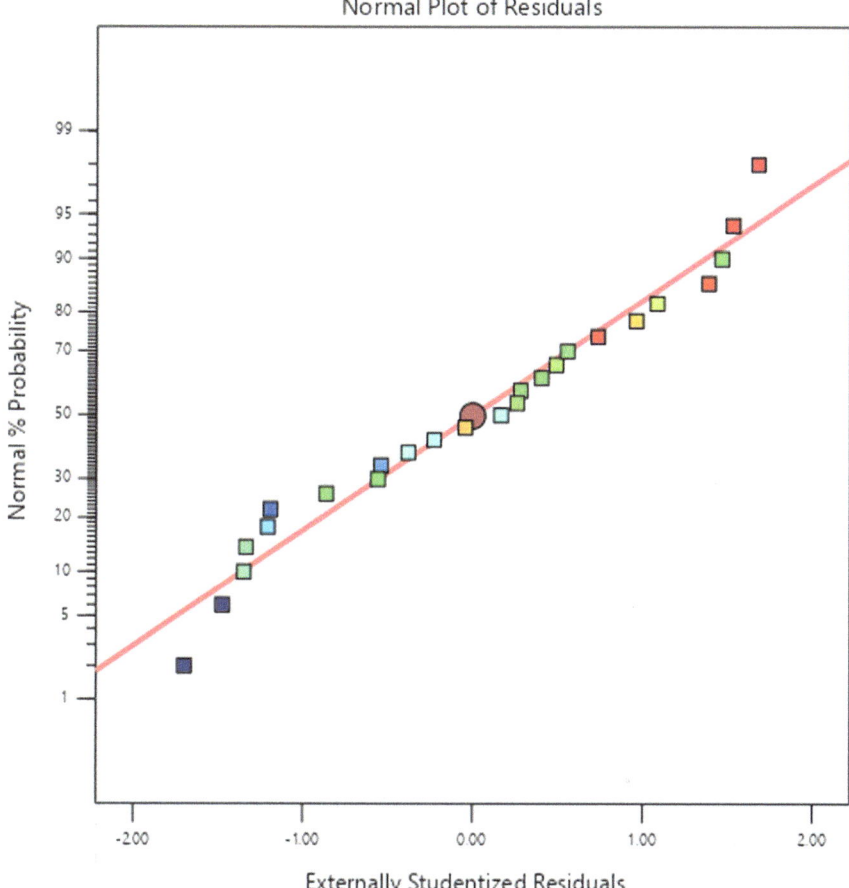

Fig. 3.32 Normal probability of residuals for μEDM milling short circuit [7]

3-Dimensional Surface Plots for MEDM Short Circuit

Figures 3.35 and 3.36 display the 3-D response surface plots produced by the model for μEDM milling short circuit. The plots illustrate the μEDM milling short circuit pattern alongside the concurrent shifts in the significant factors.

It was observed in Figs. 3.35 and 3.36 that laser power and loops have a considerable impact on the μEDM milling short circuit. The increasing trend of power decreases the μEDM milling short circuit/arching count because the high laser power removes the workpiece material to a large extent and in the μEDM milling condition less time is required to complete the milling process. Due to this reason short circuit/arching count during μEDM milling is decreased. Similarly, the loop number is increased means a more ablated zone is created in the LBMM milled channel which

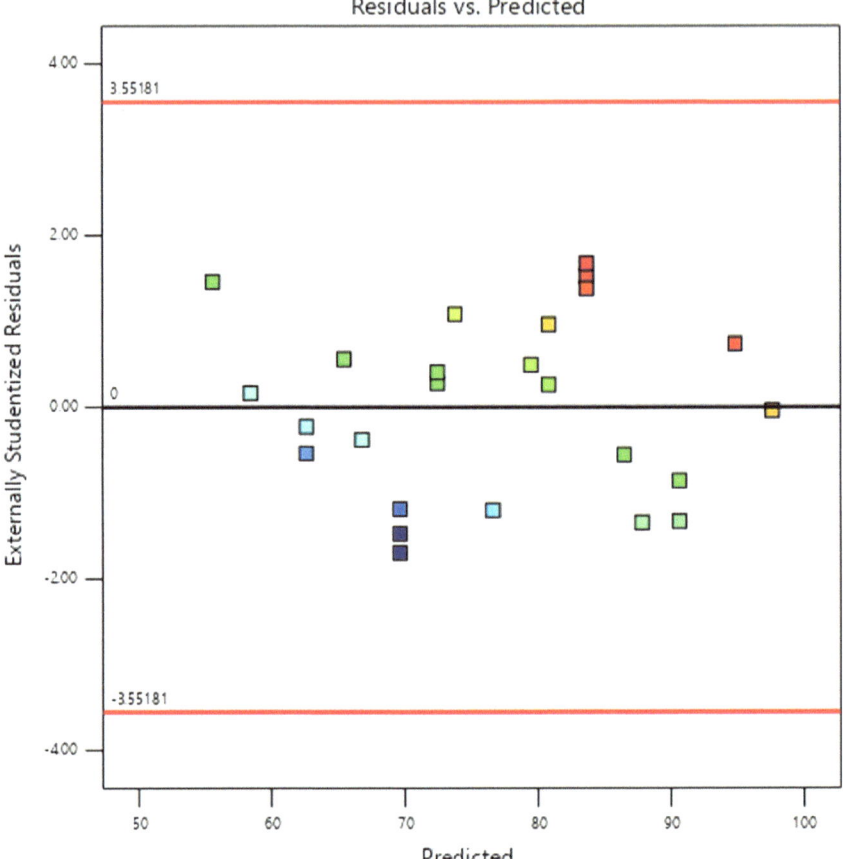

Fig. 3.33 Plot of residuals against predicted values for μEDM milling short circuit [7]

requires a shorter time to end up the μEDM milling process. It also reduces the short circuit in the μEDM milling.

Study of the Model Robustness (Short Circuit/Arching Count)

Similar to the μEDM milling time study, Fig. 3.37 was used to show the level of fitting between the experimental and predicted tool wear for the whole dataset. It was observed that 70% of the data for the short circuit was predicted within 85% of prediction accuracy which ascertains the model robustness however not as good as for the μEDM milling time.

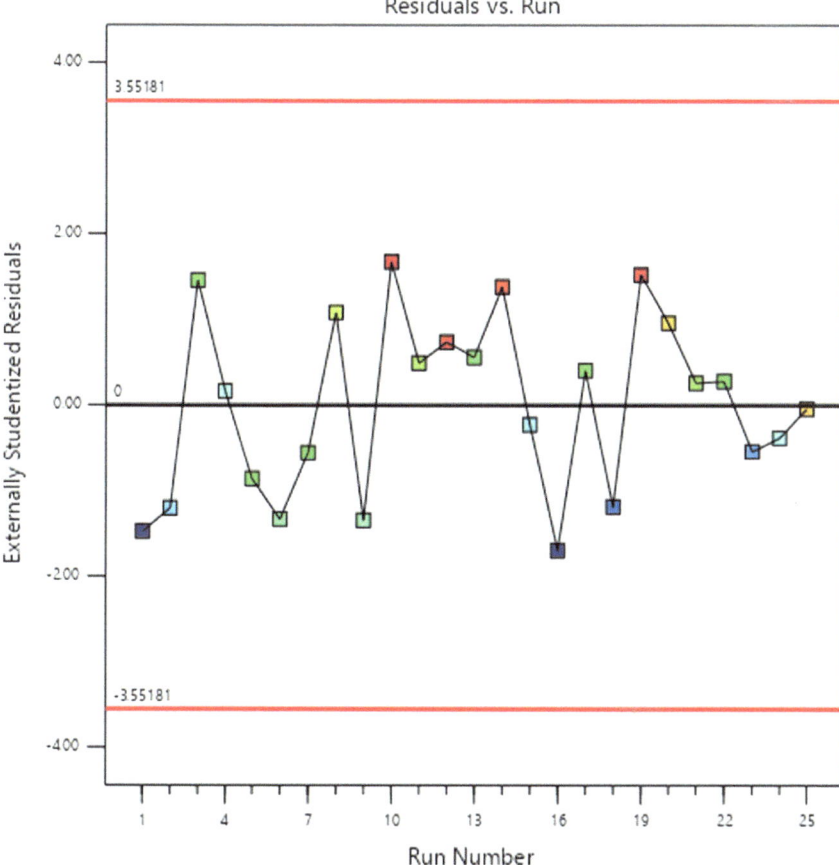

Fig. 3.34 Plot of residuals against run order for μEDM milling short circuit [7]

3.5 Summary

In Chapter 3, a dual-stage modeling approach based on artificial neural network (ANN) was described to forecast the results of the sequential process. To evaluate the output performance of LBMM-μEDM-based sequential process, in first step the laser parameters were varied and the μEDM input parameters was constant. In the second step, both the laser input parameter and the μEDM input parameters were altered in the subsequent phase of the research. The dual-stage modeling method was used, and this time the μEDM input parameters (voltage, capacitance, and EDM speed) were not kept the same. Root mean square errors (RMSEs) were calculated for each dataset and each output parameter (i.e., μEDM time, machining instability/short circuit count, and tool wear) to figure out the model accuracy. Average RMSE was calculated to be 0.050 (95% accuracy), 0.040 (96% accuracy), and 0.110 (89% accuracy) for the previously mentioned parameters. In this study's final phase, 3-D hybrid

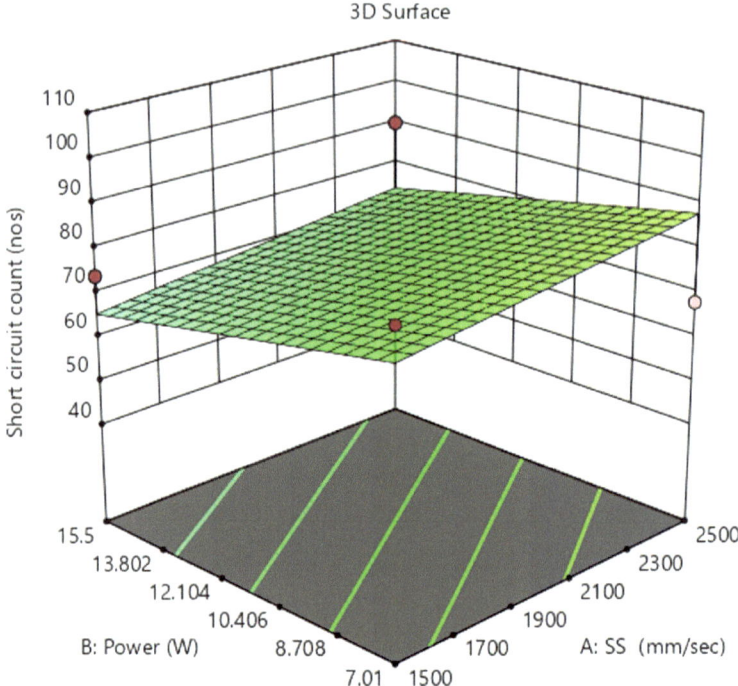

Fig. 3.35 3-D response surface plot for the μEDM milling short circuit (power and Scanning speed effect) [7]

micromachining (milling) was tested using Response Surface Methodology (RSM) to identify the significant factors influencing this sequential hybrid micromachining. It was observed that laser milling input parameters (scanning speed, power, frequency, and loops) affecting significantly on the output responses of μEDM milling time, tool wear, and machining instability (short circuit/arc count). Part of this chapter was published in reference [3].

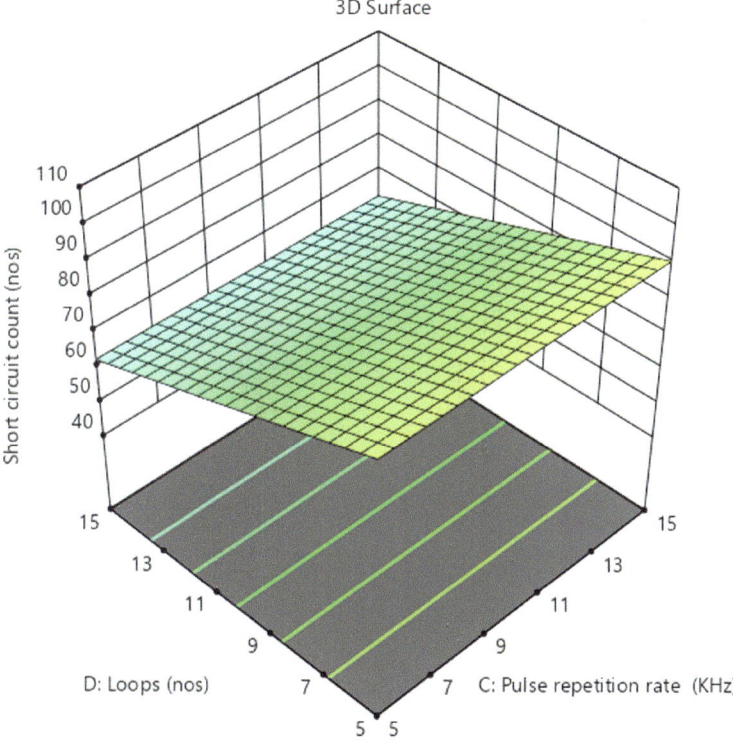

Fig. 3.36 3-D response surface plot for the μEDM milling short circuit (frequency and loop effect) [7]

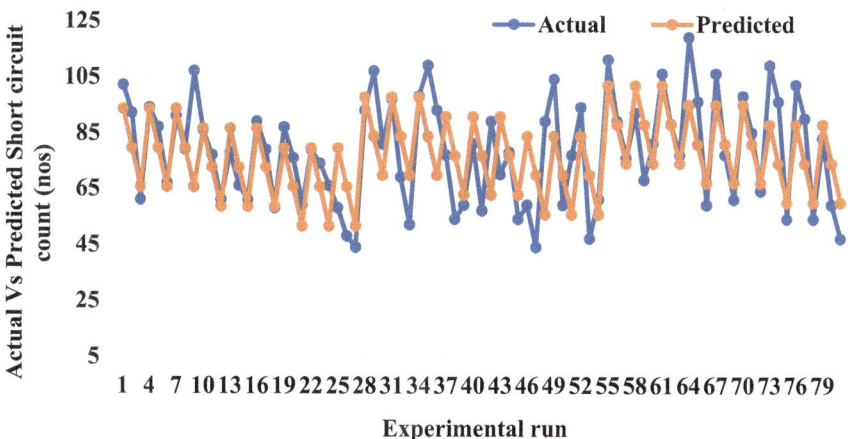

Fig. 3.37 Scattered plots of predicted against the actual values for μEDM milling short circuit [7]

Appendix 1

See Table 3.17.

Appendix 2

See Table 3.18.

Appendix 3

See Table 3.19.

Table 3.17 Dataset created from the experiments conducted during the hybrid LBMM-μEDM microdrilling with constant EDM parameters

Hole No.	LBMM input parameters			LBMM target parameters/μEDM input parameters					μEDM target parameters		
	Loop Count (nos)	Scanning speed (mm/s)	Power (W)	Pulse repetition rate (kHz)	Entry area (mm²)	Exit area (mm²)	HAZ area (mm²)	Recast area (mm²)	μEDM machining time (mins)	Short circuit/arcing count (nos)	Tool wear (μm)
1	75	50	6.4	5	0.073	0.036	0.303	0.023	18.93	87	14
2	75	50	6.4	10	0.074	0.035	0.218	0.018	18.17	49	11
3	75	50	6.4	15	0.074	0.034	0.219	0.022	18.92	41	7
4	75	50	6.4	20	0.072	0.036	0.479	0.017	17.83	47	13
5	75	50	9.4	5	0.076	0.045	0.589	0.027	15.97	46	7
6	75	50	9.4	10	0.078	0.045	0.623	0.045	17.07	39	8
7	75	50	9.4	15	0.078	0.044	0.66	0.04	15.62	43	4
8	75	50	9.4	20	0.078	0.044	0.605	0.045	17.7	75	7
9	75	50	12.5	5	0.08	0.049	0.571	0.028	17.03	39	8
10	75	50	12.5	10	0.077	0.049	0.54	0.029	14.62	41	3
11	75	50	12.5	15	0.079	0.05	0.486	0.03	16.83	43	4
12	75	50	12.5	20	0.08	0.049	0.545	0.038	16.15	44	5
13	75	50	15.5	5	0.086	0.052	0.501	0.035	14.45	32	4
14	75	50	15.5	10	0.085	0.053	0.431	0.036	14.7	37	3
15	75	50	15.5	15	0.084	0.053	0.476	0.026	15.03	33	5
16	75	50	15.5	20	0.085	0.053	0.489	0.028	15.88	46	7
17	75	500	6.4	5	0.068	0	0.217	0.011	49.25	1007	52
18	75	500	6.4	10	0.063	0	0.248	0.01	34.53	457	47

(continued)

Table 3.17 (continued)

Hole No.	LBMM input parameters				LBMM target parameters/μEDM input parameters				μEDM target parameters		
	Loop Count (nos)	Scanning speed (mm/s)	Power (W)	Pulse repetition rate (kHz)	Entry area (mm²)	Exit area (mm²)	HAZ area (mm²)	Recast area (mm²)	μEDM machining time (mins)	Short circuit/ arcing count (nos)	Tool wear (μm)
19	75	500	6.4	15	0.057	0	0.261	0.017	49.52	977	48
20	75	500	6.4	20	0.059	6.13E−05	0.247	0.014	64.87	1327	52
21	75	500	9.4	5	0.065	0.000405	0.517	0.026	32.95	401	33
22	75	500	9.4	10	0.064	0.00165	0.442	0.028	25.52	177	13
23	75	500	9.4	15	0.067	0.00122	0.525	0.016	54.5	1061	37
24	75	500	9.4	20	0.07	1.69E−05	0.477	0.016	48.23	979	42
25	75	500	12.5	5	0.072	0.014	0.725	0.025	23.28	147	15
26	75	500	12.5	10	0.061	0.01783301	0.65	0.037	22.97	141	6
27	75	500	12.5	15	0.063	0.021	0.617	0.036	19.97	105	7
28	75	500	12.5	20	0.067	0.02	0.664	0.03	19.93	79	8
29	75	500	15.5	5	0.069	0.042	0.691	0.03	16.32	33	4
30	75	500	15.5	10	0.07	0.04	0.736	0.025	17.48	78	5
31	75	500	15.5	15	0.068	0.041	0.741	0.014	16.95	50	2
32	75	500	15.5	20	0.078	0.042	0.658	0.044	17.78	64	3
33	75	950	6.4	5	0.06	0	0.246	0.014	55.07	1223	55
34	75	950	6.4	10	0.061	0	0.252	0.015	57.92	1235	53
35	75	950	6.4	15	0.064	0	0.216	0.016	46.07	902	49

(continued)

Table 3.17 (continued)

Hole No.	LBMM input parameters				LBMM target parameters/μEDM input parameters				μEDM target parameters		
	Loop Count (nos)	Scanning speed (mm/s)	Power (W)	Pulse repetition rate (kHz)	Entry area (mm²)	Exit area (mm²)	HAZ area (mm²)	Recast area (mm²)	μEDM machining time (mins)	Short circuit/ arcing count (nos)	Tool wear (μm)
36	75	950	6.4	20	0.057	0	0.254	0.015	54.25	987	39
37	75	950	9.4	5	0.067	0.00101	0.374	0.019	40.15	875	24
38	75	950	9.4	10	0.064	0.000734	0.398	0.021	48.07	966	41
39	75	950	9.4	15	0.067	0.000902	0.369	0.023	51.17	1020	44
40	75	950	9.4	20	0.066	0.000471	0.39	0.02	53.57	1091	47
41	75	950	12.5	5	0.063	0.00349	0.557	0.024	28.17	506	21
42	75	950	12.5	10	0.07	0.00549002	0.538	0.028	27.93	410	12
43	75	950	12.5	15	0.067	0.0044626	0.531	0.026	24.93	187	14
44	75	950	12.5	20	0.064	0.005546902	0.521	0.022	24.8	178	17
45	75	950	15.5	5	0.073	0.01390982	0.664	0.048	17.77	107	10
46	75	950	15.5	10	0.071	0.01466284	0.676	0.036	17.52	64	3
47	75	950	15.5	15	0.076	0.019	0.67	0.053	17.92	51	3
48	75	950	15.5	20	0.074	0.0151835	0.621	0.045	18.9	69	5
49	75	1400	6.4	5	0.067	0	0.239	0.013	50.33	986	39
50	75	1400	6.4	10	0.064	0	0.193	0.012	56.57	1107	44
51	75	1400	6.4	15	0.062	0	0.226	0.012	54.67	1226	88
52	75	1400	6.4	20	0.065	0	0.209	0.014	47.43	867	60

(continued)

Table 3.17 (continued)

Hole No.	LBMM input parameters				LBMM target parameters/μEDM input parameters				μEDM target parameters		
	Loop Count (nos)	Scanning speed (mm/s)	Power (W)	Pulse repetition rate (kHz)	Entry area (mm²)	Exit area (mm²)	HAZ area (mm²)	Recast area (mm²)	μEDM machining time (mins)	Short circuit/ arcing count (nos)	Tool wear (μm)
53	75	1400	9.4	5	0.068	0	0.383	0.016	47.23	1064	31
54	75	1400	9.4	10	0.068	0	0.305	0.014	36.1	781	28
55	75	1400	9.4	15	0.064	0	0.357	0.014	38.67	787	27
56	75	1400	9.4	20	0.061	0	0.357	0.013	46.7	988	51
57	75	1400	12.5	5	0.069	0.00259842	0.502	0.013	40.37	823	38
58	75	1400	12.5	10	0.07	0.00205	0.466	0.017	38.28	771	33
59	75	1400	12.5	15	0.067	0.00379302	0.485	0.014	32.08	635	31
60	75	1400	12.5	20	0.067	0.00103	0.435	0.018	37.68	793	49
61	75	1400	15.5	5	0.068	0.00942211	0.629	0.023	22	97	8
62	75	1400	15.5	10	0.075	0.013	0.627	0.038	16.32	57	10
63	75	1400	15.5	15	0.07	0.01044373	0.594	0.032	16.6	61	11
64	75	1400	15.5	20	0.067	0.010813	0.685	0.03	16.12	67	13
65	25	1400	6.4	10	0.06	0	0.227	0.016	55.05	1155	68
66	50	1400	6.4	10	0.063	0	0.208	0.017	57.85	1104	74
67	75	1400	6.4	10	0.068	0	0.236	0.016	47.43	1012	69
68	75	1400	6.4	10	0.068	0	0.263	0.015	63.48	1276	81
69	100	1400	6.4	10	0.064	0	0.265	0.016	62.52	1488	86

(continued)

Table 3.17 (continued)

Hole No.	LBMM input parameters					LBMM target parameters/μEDM input parameters					μEDM target parameters		
	Loop Count (nos)	Scanning speed (mm/s)	Power (W)	Pulse repetition rate (kHz)		Entry area (mm²)	Exit area (mm²)	HAZ area (mm²)	Recast area (mm²)		μEDM machining time (mins)	Short circuit/arcing count (nos)	Tool wear (μm)
70	25	50	15.5	10		0.087	0.047	0.392	0.046		18.95	89	7
71	50	50	15.5	10		0.082	0.051	0.463	0.045		16.67	37	2
72	75	50	15.5	10		0.089	0.053	0.408	0.045		17.83	43	4
73	75	50	15.5	10		0.081	0.053	0.467	0.041		17.78	77	5
74	100	50	15.5	10		0.081	0.053	0.408	0.042		19.17	83	5
75	25	400	10.7	10		0.066	0.00021015	0.355	0.019		39.45	689	31
76	25	400	11.3	10		0.067	0.00027	0.375	0.021		40.25	736	38
77	25	400	11.9	10		0.068	0.00040441	0.405	0.025		36.6	580	28
78	25	600	10.7	10		0.063	0	0.317	0.021		40.45	671	33
79	25	600	11.9	10		0.062	6.22E−05	0.357	0.022		39.83	721	36
80	25	600	13.1	10		0.064	0.00025506	0.422	0.021		45.15	782	33
81	25	700	12.5	10		0.063	0	0.369	0.025		52.18	897	49
82	25	700	13.1	10		0.064	0	0.386	0.022		49.08	816	42
83	25	700	13.7	10		0.064	0	0.427	0.022		59.17	1172	59
84	25	900	10.7	10		0.06	2.40E−05	0.343	0.019		60.33	1241	54
85	25	900	11.3	10		0.06	0	0.326	0.02		43.02	758	35
86	25	900	11.9	10		0.061	8.80E−05	0.349	0.02		49.35	981	41

(continued)

Table 3.17 (continued)

Hole No.	LBMM input parameters			LBMM target parameters/μEDM input parameters					μEDM target parameters		
	Loop Count (nos)	Scanning speed (mm/s)	Power (W)	Pulse repetition rate (kHz)	Entry area (mm²)	Exit area (mm²)	HAZ area (mm²)	Recast area (mm²)	μEDM machining time (mins)	Short circuit/arcing count (nos)	Tool wear (μm)
87	25	1400	13.1	10	0.062	0.000471	0.401	0.019	48.85	966	43
88	25	1400	14.3	10	0.064	0.00079012	0.426	0.022	45.05	807	44
89	25	1400	14.9	10	0.065	0.0010562	0.469	0.023	43.38	765	39
90	50	400	10.7	10	0.061	0.00073279	0.434	0.019	37.65	691	28
91	50	400	11.3	10	0.067	0.00197965	0.448	0.029	33.97	511	25
92	50	400	11.9	10	0.06	0.004206834	0.523	0.019	26.2	251	18
93	50	600	10.7	10	0.064	0.00050653	0.391	0.023	62.02	1156	52
94	50	600	11.9	10	0.064	0.000686951	0.398	0.024	26.07	211	17
95	50	600	13.1	10	0.063	0.001507734	0.504	0.021	25.22	188	13
96	50	700	12.5	10	0.063	0.001151301	0.487	0.025	38.77	587	27
97	50	700	13.1	10	0.065	0.00184682	0.462	0.018	26.02	222	16
98	50	700	13.7	10	0.068	0.001691439	0.514	0.026	27.73	230	14
99	50	900	10.7	10	0.063	0.000268868	0.366	0.018	44.97	788	39
100	50	900	11.3	10	0.064	0.000273766	0.356	0.022	52.18	941	53
101	50	900	11.9	10	0.063	0.00046,654	0.394	0.021	51.28	1022	47
102	50	1400	13.1	10	0.062	0.001237941	0.437	0.022	44.8	831	37
103	50	1400	14.3	10	0.065	0.00206935	0.499	0.021	34.53	593	29

(continued)

Table 3.17 (continued)

Hole No.	LBMM input parameters				LBMM target parameters/μEDM input parameters				μEDM target parameters		
	Loop Count (nos)	Scanning speed (mm/s)	Power (W)	Pulse repetition rate (kHz)	Entry area (mm²)	Exit area (mm²)	HAZ area (mm²)	Recast area (mm²)	μEDM machining time (mins)	Short circuit/arcing count (nos)	Tool wear (μm)
104	50	1400	14.9	10	0.066	0.00278	0.518	0.02	23.88	223	16
105	100	400	10.7	10	0.059	0.01907776	0.711	0.016	24.4	167	14
106	100	400	11.3	10	0.059	0.021	0.465	0.022	14.47	59	4
107	100	400	11.9	10	0.068	0.03	0.455	0.022	15.63	62	7
108	100	600	10.7	10	0.059	0.0081604	0.557	0.022	17.97	119	5
109	100	600	11.9	10	0.057	0.0110431	0.545	0.018	16.33	105	6
110	100	600	13.1	10	0.059	0.0194692	0.617	0.015	18.27	87	5
111	100	700	12.5	10	0.058	0.011267956	0.551	0.024	28.62	337	21
112	100	700	13.1	10	0.065	0.01437727	0.611	0.024	16.9	59	5
113	100	700	13.7	10	0.064	0.019186627	0.657	0.032	33.75	491	23
114	100	900	10.7	10	0.064	0.000954463	0.47	0.022	24.75	233	18
115	100	900	11.3	10	0.063	0.00162465	0.492	0.019	37.42	624	24
116	100	900	11.9	10	0.062	0.002178024	0.544	0.022	27.12	267	15
117	100	1400	13.1	10	0.063	0.00265903	0.542	0.023	31.07	527	16
118	100	1400	14.3	10	0.064	0.00416066	0.582	0.023	25.53	257	14
119	100	1400	14.9	10	0.066	0.00790337	0.639	0.026	24.13	184	15

Table 3.18 Dataset created from the experiments conducted during the hybrid LBMM-μEDM drilling with variable EDM parameter

Hole No.	LBMM-uEDM input parameters						LBMM output parameters			μEDM target parameters		
	Loop count (nos)	Scanning speed (mm/s)	Power (W)	Pulse repetition rate (kHz)	Discharge energy (nJ)	EDM speed (um/s)	Volume removed (mm³)	HAZ Area (mm²)	Recast area (mm²)	μEDM machining time (s)	Short circuit/arcing count (nos)	Tool wear (μm)
1	75	1400	6.4	20	605,000	10	0.003006533	0.239	0.004	776.5	23	29
2	75	1400	6.4	5	320,000	10	0.003157095	0.224	0.002	838.5	34	31
3	75	1400	15.5	20	605,000	3	0.00767702	0.477	0.004	831.5	2	20
4	75	1400	15.5	5	605,000	3	0.006686593	0.45	0.009	849.5	3	21
5	75	725	11	12.5	4512.5	6.5	0.006399907	0.356	0.01	1602.87	311	19
6	75	1400	6.4	20	320,000	3	0.003247099	0.224	0.003	791.5	3	22
7	75	725	8.7	12.5	4512.5	6.5	0.004732751	0.316	0.008	2110.87	667	26
8	75	725	11	12.5	4512.5	6.5	0.006122056	0.392	0.005	1901.87	589	23
9	75	50	15.5	5	60.5	3	0.032001759	0.44	0.023	1837.81	611	9
10	75	50	15.5	5	320,000	10	0.024395855	0.387	0.023	931.81	3	21
11	75	50	6.4	5	32	10	0.024407595	0.409	0.004	2521.81	827	12
12	75	50	15.5	5	60.5	10	0.031157148	0.423	0.032	2375.81	781	11
13	75	50	15.5	20	605,000	10	0.031373323	0.439	0.024	516.81	6	16
14	75	50	6.4	5	605,000	10	0.026088964	0.464	0.012	583.81	3	23
15	75	50	15.5	20	605,000	3	0.031966157	0.447	0.02	877.81	3	20
16	75	725	13.23	12.5	4512.5	6.5	0.005695416	0.452	0.003	1427.87	107	12
17	75	1400	15.5	5	32	3	0.006812039	0.479	0.006	3914.5	1376	41
18	75	725	11	12.5	4512.5	6.5	0.00565909	0.374	0.008	1814.87	504	26

(continued)

Table 3.18 (continued)

Hole No.	LBMM-uEDM input parameters						LBMM output parameters			µEDM target parameters		
	Loop count (nos)	Scanning speed (mm/s)	Power (W)	Pulse repetition rate (kHz)	Discharge energy (nJ)	EDM speed (um/s)	Volume removed (mm^3)	HAZ Area (mm^2)	Recast area (mm^2)	µEDM machining time (s)	Short circuit/ arcing count (nos)	Tool wear (µm)
19	75	50	6.4	20	320,000	10	0.025186688	0.449	0.009	903.81	16	21
20	75	1400	6.4	5	60.5	3	0.003240284	0.221	0.003	3941.5	1231	13
21	75	725	11	12.5	3828.125	6.5	0.006777124	0.372	0.008	2056.87	533	26
22	75	725	11	12.5	4512.5	6.5	0.004325694	0.4	0.0028727	1972.87	491	21
23	75	50	6.4	20	32	10	0.025576016	0.459	0.007	4061.81	1384	15
24	75	1400	15.5	5	605,000	10	0.004914559	0.509	0.007	642.5	1	18
25	75	725	11	12.5	4512.5	6.5	0.00494093	0.391	0.004	1733.87	563	23
26	75	387.5	11	12.5	4512.5	6.5	0.016778292	0.424	0.008	1084.45	87	7
27	75	725	11	12.5	4512.5	6.5	0.004187553	0.37	0.004	1400.87	341	13
28	75	725	11	12.5	45,125	6.5	0.003801182	0.354	0.004	861.87	47	19
29	75	725	11	8.75	4512.5	6.5	0.003784449	0.379	0.005	2168.87	534	29
30	75	725	11	16.25	4512.5	6.5	0.004125375	0.37	0.005	2086.87	476	22
31	75	725	11	12.5	4512.5	8.25	0.004224455	0.386	0.005	2100.87	653	26
32	75	1400	6.4	5	605,000	10	0.002107243	0.228	0.002	599.5	2	19
33	75	1400	6.4	5	32	10	0.000727828	0.228	0.002	7810.5	2684	16
34	75	725	11	12.5	4512.5	6.5	0.003805643	0.37	0.008	2251.87	587	31
35	75	50	6.4	20	60.5	10	0.025703501	0.463	0.006	3312.81	1187	12

(continued)

Table 3.18 (continued)

Hole No.	LBMM-uEDM input parameters						LBMM output parameters			μEDM target parameters		
	Loop count (nos)	Scanning speed (mm/s)	Power (W)	Pulse repetition rate (kHz)	Discharge energy (nJ)	EDM speed (um/s)	Volume removed (mm³)	HAZ Area (mm²)	Recast area (mm²)	μEDM machining time (s)	Short circuit/ arcing count (nos)	Tool wear (μm)
36	75	1400	6.4	5	320,000	3	0.00278891	0.244	0.001	870.5	7	22
37	75	1400	15.5	20	60.5	10	0.005320032	0.468	0.003	5055.5	1673	18
38	75	50	15.5	5	32	10	0.031283274	0.38	0.015	4404.81	1491	13
39	75	1400	6.4	5	32	3	0.002701387	0.236	0.002	9748.5	3981	24
40	75	50	6.4	5	32	3	0.025190555	0.361	0.006	4756.81	1577	16
41	75	1400	15.5	20	32	3	0.007884011	0.379	0.005	6281.5	2289	17
42	75	1400	15.5	20	32	10	0.007426965	0.374	0.002	5418.5	1959	19
43	75	50	6.4	5	320,000	3	0.025460416	0.36	0.012	738.5	5	18
44	75	1400	15.5	5	320,000	3	0.007520275	0.338	0.011	1326.5	9	27
45	75	50	15.5	20	60.5	3	0.031189779	0.361	0.033	3710.81	1380	17
46	75	50	6.4	20	320,000	3	0.026850199	0.366	0.006	818.81	4	18
47	75	50	6.4	20	32	3	0.027100968	0.314	0.003	2912.81	945	5
48	75	1400	15.5	20	60.5	3	0.007894187	0.424	0.005	2573.5	841	8
49	75	50	15.5	5	605,000	10	0.02998891	0.334	0.028	692.81	5	11
50	75	50	6.4	5	60.5	10	0.026193798	0.348	0.015	3638.81	1362	13
51	75	50	6.4	20	605,000	10	0.02709235	0.327	0.016	734.81	10	17
52	75	1400	15.5	20	320,000	10	0.008525444	0.264	0.009	975.5	17	24

(continued)

Table 3.18 (continued)

Hole No.	LBMM-uEDM input parameters						LBMM output parameters			µEDM target parameters		
	Loop count (nos)	Scanning speed (mm/s)	Power (W)	Pulse repetition rate (kHz)	Discharge energy (nJ)	EDM speed (um/s)	Volume removed (mm³)	HAZ Area (mm²)	Recast area (mm²)	µEDM machining time (s)	Short circuit/ arcing count (nos)	Tool wear (µm)
53	75	725	11	12.5	4512.5	6.5	0.005744009	0.254	0.004	2435.87	786	31
54	75	1400	6.4	20	32	3	0.003133525	0.128	0.003	8315.5	3398	16
55	75	50	15.5	20	60.5	10	0.030493527	0.343	0.014	2135.81	688	8
56	75	50	6.4	5	320,000	10	0.024920917	0.235	0.005	961.81	13	17
57	75	1400	15.5	5	60.5	3	0.008537747	0.34	0.012	5350.5	1851	17
58	75	50	15.5	5	32	3	0.029956058	0.289	0.023	4096.81	1327	11
59	75	1400	6.4	20	320,000	10	0.002912424	0.136	0.003	1310.5	12	28
60	75	1400	15.5	20	320,000	3	0.009571608	0.27	0.011	1534.5	13	13
61	75	1400	15.5	5	60.5	10	0.009955243	0.277	0.008	3783.5	1279	15
62	75	1400	6.4	20	60.5	3	0.002381445	0.133	0.002	5472.5	1876	17
63	75	1400	6.4	5	60.5	10	0.002555999	0.105	0.003	4265.5	1421	13
64	75	50	15.5	5	320,000	3	0.029890106	0.301	0.023	814.81	11	19
65	75	725	11	12.5	4512.5	6.5	0.007276511	0.18	0.006	1251.87	156	11
66	75	50	6.4	20	60.5	3	0.024541304	0.174	0.008	2288.81	731	7
67	75	1400	6.4	5	605,000	3	0.003044012	0.116	0.003	1153.5	17	21
68	75	50	15.5	20	32	10	0.029535677	0.266	0.026	4158.81	1326	14
					60.5	10	0.003246534	0.171	0.003	6768.5	2498	17

(continued)

Table 3.18 (continued)

Hole No.	LBMM-uEDM input parameters						LBMM output parameters			µEDM target parameters		
	Loop count (nos)	Scanning speed (mm/s)	Power (W)	Pulse repetition rate (kHz)	Discharge energy (nJ)	EDM speed (um/s)	Volume removed (mm³)	HAZ Area (mm²)	Recast area (mm²)	µEDM machining time (s)	Short circuit/arcing count (nos)	Tool wear (µm)
69	75	1400	6.4	20	605,000	3	0.024267759	0.196	0.01	980.81	11	19
70	75	50	6.4	5	32	10	0.008061082	0.366	0.012	10,119.5	3631	21
71	75	1400	15.5	5	605,000	10	0.008094203	0.389	0.01	806.5	5	12
72	75	1400	15.5	20	320,000	10	0.008192221	0.375	0.012	885.5	9	7
73	75	1400	15.5	5	4512.5	6.5	0.005393862	0.328	0.005	2192.63	688	27
74	75	1062.5	11	12.5	32	10	0.003009605	0.197	0.003	8683.5	3459	17
75	75	1400	6.4	20	605,000	3	0.031625801	0.309	0.022	764.81	3	18
76	75	50	15.5	5	320,000	3	0.030477131	0.303	0.008	821.81	3	19
77	75	50	15.5	20	605,000	3	0.026317711	0.455	0.014	953.81	6	21
78	75	50	6.4	20	60.5	3	0.026342443	0.44	0.003	6113.81	2641	16
79	75	50	6.4	5	4512.5	6.5	0.007611589	0.364	0.008	1765.87	411	23
80	75	725	11	12.5	32	3	0.031587114	0.401	0.027	4453.81	1658	16
81	75	50	15.5	20	320,000	10	0.030776444	0.401	0.025	766.81	3	16
82	75	50	15.5	20	4512.5	4.75	0.007621099	0.365	0.007	1796.87	469	24
83	75	725	11	12.5	5253.125	6.5	0.007091388	0.352	0.002	1389.87	387	17
84	75	725	11	12.5	605,000	3	0.002614485	0.235	0.003	1019.5	2	16
85	75	1400	6.4	20	451.25	6.5	0.007624468	0.399	0.003	3852.87	1516	33

(continued)

Table 3.18 (continued)

Hole No.	LBMM-uEDM input parameters						LBMM output parameters			μEDM target parameters		
	Loop count (nos)	Scanning speed (mm/s)	Power (W)	Pulse repetition rate (kHz)	Discharge energy (nJ)	EDM speed (um/s)	Volume removed (mm^3)	HAZ Area (mm^2)	Recast area (mm^2)	μEDM machining time (s)	Short circuit/ arcing count (nos)	Tool wear (μm)
86	75	725	11	12.5	4512.5	6.5	0.007708557	0.382	0.004	1400.87	439	13
87	75	725	11	12.5	4512.5	8.25	0.004305037	0.285	0.006	1667.63	511	16
88	75	1062.5	8.7	12.5	45,125	6.5	0.008466447	0.442	0.007	646.87	26	15
89	75	725	13.2	16.25	45,125	4.75	0.005836636	0.347	0.01	760.63	25	17
90	75	1062.5	11	12.5	4512.5	4.75	0.02604787	0.479	0.031	1296.45	87	7
91	75	387.5	13.2	12.5	52,531.25	6.5	0.007770149	0.479	0.021	591.87	19	11
92	75	725	13.2	12.5	4512.5	6.5	0.007359969	0.363	0.007	1062.87	185	9
93	75	725	11	12.5	5253.125	8.25	0.0074411598	0.369	0.005	980.87	176	9
94	75	725	11	16.25	3828.125	8.25	0.007209127	0.382	0.01	1260.87	306	19
95	75	725	11	8.75	5253.125	6.5	0.018785684	0.441	0.007	911.45	81	10
96	75	387.5	11	8.75	3828.125	6.5	0.019489454	0.439	0.007	1202.45	161	18
97	75	387.5	11	16.25	5253.125	6.5	0.006186077	0.342	0.009	1736.63	448	24
98	75	1062.5	11	16.25	4512.5	8.25	0.026503302	0.559	0.033	1073.45	97	8
99	75	387.5	13.2	12.5	5253.125	6.5	0.019558828	0.418	0.008	1197.45	102	5
100	75	387.5	11	16.25	3828.125	8.25	0.008570618	0.488	0.008	1647.87	251	15
101	75	725	11	16.25	3828.125	4.75	0.009137857	0.465	0.008	1665.87	261	10
102	75	725	11	8.75	5253.125	8.25	0.008793497	0.439	0.006	1358.87	182	7

(continued)

Table 3.18 (continued)

Hole No.	LBMM-uEDM input parameters						LBMM output parameters			µEDM target parameters		
	Loop count (nos)	Scanning speed (mm/s)	Power (W)	Pulse repetition rate (kHz)	Discharge energy (nJ)	EDM speed (um/s)	Volume removed (mm³)	HAZ Area (mm²)	Recast area (mm²)	µEDM machining time (s)	Short circuit/ arcing count (nos)	Tool wear (µm)
103	75	725	11	8.75	4512.5	4.75	0.008536005	0.383	0.007	1497.45	219	11
104	75	387.5	8.7	12.5	3828.125	6.5	0.005346869	0.403	0.003	3617.63	988	40
105	75	1062.5	11	16.25	525.3125	6.5	0.008938238	0.474	0.021	3709.87	801	18
106	75	725	13.2	12.5	4512.5	4.75	0.003676122	0.348	0.003	2757.63	689	28
107	75	1062.5	8.7	12.5	52,531.25	6.5	0.005916394	0.398	0.005	733.87	27	14
108	75	725	8.7	12.5	3828.125	6.5	0.022874903	0.53	0.022	972.45	111	6
109	75	387.5	11	8.75	5253.125	4.75	0.008361408	0.444	0.004	1349.87	327	24
110	75	725	11	16.25	45,125	6.5	0.010302736	0.473	0.03	655.87	31	7
111	75	725	11	8.75	4512.5	6.5	0.00831561	0.419	0.01	1241.87	311	11
112	75	725	11	12.5	451.25	8.25	0.022885096	0.495	0.031	4417.45	1338	46
113	75	387.5	11	12.5	4512.5	6.5	0.007853675	0.418	0.01	1309.87	197	11
114	75	725	11	12.5	451.25	6.5	0.005387531	0.348	0.006	2411.87	616	27
115	75	725	8.7	16.25	4512.5	6.5	0.008689374	0.44	0.013	1210.87	187	11
116	75	725	11	12.5	4512.5	6.5	0.008249746	0.44	0.006	1937.87	513	19
117	75	725	11	12.5	3828.125	4.75	0.008424768	0.417	0.012	1631.87	387	18
118	75	725	11	16.25	45,125	6.5	0.006030265	0.342	0.005	746.87	37	12
119	75	725	8.7	16.25	4512.5	4.75	0.009523174	0.456	0.012	865.63	81	6

(continued)

Table 3.18 (continued)

Hole No.	LBMM-uEDM input parameters						LBMM output parameters			µEDM target parameters		
	Loop count (nos)	Scanning speed (mm/s)	Power (W)	Pulse repetition rate (kHz)	Discharge energy (nJ)	EDM speed (um/s)	Volume removed (mm³)	HAZ Area (mm²)	Recast area (mm²)	µEDM machining time (s)	Short circuit/arcing count (nos)	Tool wear (µm)
120	75	1062.5	13.2	12.5	451.25	4.75	0.00673978	0.368	0.006	5026.63	1838	61
121	75	1062.5	11	12.5	5253.125	4.75	0.008166229	0.433	0.01	776.87	76	5
122	75	725	11	8.75	451.25	6.5	0.005904369	0.333	0.004	6294.87	2166	67
123	75	725	8.7	8.75	382.8125	6.5	0.01015753	0.461	0.021	4215.87	1218	44
124	75	725	13.2	12.5	45,125	8.25	0.006453581	0.385	0.007	716.63	12	13
125	75	1062.5	11	12.5	3828.125	6.5	0.006039508	0.389	0.006	1553.63	537	17
126	75	1062.5	11	8.75	451.25	8.25	0.006353226	0.396	0.007	4645.63	1697	48
127	75	1062.5	11	12.5	45,125	4.75	0.008773956	0.443	0.006	666.45	20	19
128	75	387.5	11	12.5	45,125	6.5	0.005750128	0.349	0.008	892.87	24	23
129	75	725	8.7	8.75	451.25	6.5	0.008497277	0.46	0.027	5500.87	1761	59
130	75	725	13.2	8.75	4512.5	8.25	0.009211673	0.444	0.016	1768.63	468	17
131	75	1062.5	13.2	12.5	5253.125	6.5	0.00646141	0.413	0.006	1578.63	411	14
132	75	1062.5	11	8.75	4512.5	8.25	0.008067021	0.349	0.008	1005.45	328	9
133	75	387.5	8.7	12.5	45,125	8.25	0.011703391	0.42	0.015	4961.45	1689	54
134	75	387.5	11	12.5	451.25	6.5	0.009120124	0.428	0.015	5483.87	1561	48
135	75	725	13.2	16.25	38,281.25	6.5	0.008866852	0.477	0.019	821.87	19	18
136	75	725	13.2	12.5	525.3125	6.5	0.0058252	0.355	0.005	5022.87	1497	49

(continued)

Table 3.18 (continued)

Hole No.	LBMM-uEDM input parameters							LBMM output parameters				µEDM target parameters		
	Loop count (nos)	Scanning speed (mm/s)	Power (W)	Pulse repetition rate (kHz)	Discharge energy (nJ)	EDM speed (um/s)		Volume removed (mm³)	HAZ Area (mm²)	Recast area (mm²)		µEDM machining time (s)	Short circuit/ arcing count (nos)	Tool wear (µm)
137	75	725	8.7	12.5	38,281.25	6.5		0.006175429	0.328	0.004		777.87	21	26
138	75	725	8.7	12.5	382.8125	6.5		0.005904856	0.323	0.006		6387.87	19,017	55
139	75	725	8.7	12.5	451.25	4.75		0.009477535	0.431	0.021		5156.45	1288	41
140	75	387.5	11	12.5	605,000	10		0.003006533	0.239	0.004		776.5	23	29

Table 3.19 Dataset created from the experiments conducted during the hybrid LBMM-μEDM micromachining (milling) with constant EDM parameters

	LBMM input parameters				μEDM output parameters (Actual)		
No.	Loop (nos)	Scanning speed D (mm/s)	Power percentage (%)	Pulse repetition rate	uEDM time (min)	Tool wear (μm)	Short circuits (nos)
1	5	1500	20	5	53.2	36	102
2	10	1500	20	5	49.25	30	92
3	15	1500	20	5	70.28	18	61
4	5	1500	20	10	59.02	33	94
5	10	1500	20	10	45.17	31	87
6	15	1500	20	10	49.26	29	67
7	5	1500	20	15	51.35	37	91
8	10	1500	20	15	50.56	33	79
9	15	1500	20	15	53.48	39	107
10	5	1500	55	5	48.31	31	86
11	10	1500	55	5	44.56	28	77
12	15	1500	55	5	41.21	23	61
13	5	1500	55	10	41.23	29	78
14	10	1500	55	10	39.08	26	66
15	15	1500	55	10	38.08	19	61
16	5	1500	55	15	62.37	39	89
17	10	1500	55	15	42.54	33	79
18	15	1500	55	15	40.31	30	58
19	5	1500	90	5	44.56	33	87
20	10	1500	90	5	39.56	28	76
21	15	1500	90	5	32.48	29	59
22	5	1500	90	10	36.51	33	79
23	10	1500	90	10	38.22	31	74
24	15	1500	90	10	32.08	25	66
25	5	1500	90	15	31.59	27	58
26	10	1500	90	15	32.16	23	48
27	15	1500	90	15	29.07	18	44
28	5	2000	20	5	48.31	34	93
29	10	2000	20	5	52.3	41	107
30	15	2000	20	5	46.29	36	81
31	5	2000	20	10	48.41	39	97
32	10	2000	20	10	44.59	31	69
33	15	2000	20	10	43.22	24	52
34	5	2000	20	15	48.44	32	98

(continued)

Table 3.19 (continued)

No.	LBMM input parameters				μEDM output parameters (Actual)		
	Loop (nos)	Scanning speed D (mm/s)	Power percentage (%)	Pulse repetition rate	uEDM time (min)	Tool wear (μm)	Short circuits (nos)
35	10	2000	20	15	48.14	33	109
36	15	2000	20	15	45.37	28	93
37	5	2000	55	5	38.56	26	77
38	10	2000	55	5	38.08	21	54
39	15	2000	55	5	38.37	23	59
40	5	2000	55	10	46.57	34	81
41	10	2000	55	10	41.36	31	57
42	15	2000	55	10	43.59	30	89
43	5	2000	55	15	44.52	28	70
44	10	2000	55	15	42.06	26	78
45	15	2000	55	15	36.51	26	54
46	5	2000	90	5	34.5	28	59
47	10	2000	90	5	34.02	23	44
48	15	2000	90	5	42.11	31	89
49	5	2000	90	10	36.22	29	104
50	10	2000	90	10	32.55	21	59
51	15	2000	90	10	34.21	26	77
52	5	2000	90	15	37.41	31	94
53	10	2000	90	15	30.58	29	47
54	15	2000	90	15	30.22	21	61
55	5	2500	20	5	52.57	41	111
56	10	2500	20	5	51.54	35	89
57	15	2500	20	5	52.48	31	76
58	5	2500	20	10	48.21	38	92
59	10	2500	20	10	47.03	27	68
60	15	2500	20	10	47.19	31	81
61	5	2500	20	15	51.59	33	106
62	10	2500	20	15	49.3	30	88
63	15	2500	20	15	48.51	26	77
64	5	2500	55	5	44.31	34	119
65	10	2500	55	5	44.09	31	96
66	15	2500	55	5	41.01	22	59
67	5	2500	55	10	46.5	31	106

(continued)

Table 3.19 (continued)

	LBMM input parameters				μEDM output parameters (Actual)		
No.	Loop (nos)	Scanning speed D (mm/s)	Power percentage (%)	Pulse repetition rate	uEDM time (min)	Tool wear (μm)	Short circuits (nos)
68	10	2500	55	10	38.32	24	77
69	15	2500	55	10	37.05	24	61
70	5	2500	55	15	45.11	31	98
71	10	2500	55	15	48.43	36	85
72	15	2500	55	15	41.49	29	64
73	5	2500	90	5	37.49	28	109
74	10	2500	90	5	38.39	28	96
75	15	2500	90	5	34.11	23	54
76	5	2500	90	10	41.2	31	102
77	10	2500	90	10	38.51	26	90
78	15	2500	90	10	35.55	24	54
79	5	2500	90	15	38.49	29	83
80	10	2500	90	15	36.44	25	59
81	15	2500	90	15	34.56	21	47

References

1. Kumar S, Batish A, Singh R, Singh TP (2014) A hybrid Taguchi-artificial neural network approach to predict surface roughness during electric discharge machining of titanium alloys. J Mech Sci Technol 28(7):2831–2844. https://doi.org/10.1007/s12206-014-0637-x
2. Parandoush P, Hossain A (2014) A review of modeling and simulation of laser beam machining. Int J Mach Tools Manuf 85:135–145. https://doi.org/10.1016/j.ijmachtools.2014.05.008
3. Noor WI, Saleh T, Rashid MAN, Mohd Ibrahim A, Ali MSM (2021) Dual-stage artificial neural network (ANN) model for sequential LBMM-μEDM-based micro-drilling. Int J Adv Manuf Technol 117(11–12):3343–3365. https://doi.org/10.1007/s00170-021-07910-w
4. Bre F, Gimenez JM, Fachinotti VD (2018) Prediction of wind pressure coefficients on building surfaces using artificial neural networks. Energy Build 158:1429–1441. https://doi.org/10.1016/j.enbuild.2017.11.045
5. Burden F, Winkler D (2008) Bayesian regularization of neural networks. Methods Mol Biol 458:25–44. https://doi.org/10.1007/978-1-60327-101-1_3
6. Noor WI (2023) ANN modeling of laser-micro electro discharge machining based hybrid microdrilling process. International Islamic University Malaysia
7. Rashid MAN (2024) Investigation on laser beam and micro electro discharge machining based hybrid micromachining. International Islamic University Malaysia
8. Astakhov VP (2012) Design of experiment methods in manufacturing: Basics and practical applications. Statist Comput Techniq Manuf 9783642258:1–54. https://doi.org/10.1007/978-3-642-25859-6_1
9. Mullen K, Hultquist RA (2017) Introduction to statistics. Biometrics 26(3):590. https://doi.org/10.2307/2529118
10. Pike ER, McNally B (1997) Theory and design of photon correlation and light-scattering experiments 36(30). https://doi.org/10.1364/ao.36.007531

11. Montgomery D (2009) Design_Mont_Part1.Pdf
12. Rashid MAN, Saleh T, Abdul Hamid SB, Rashid MM (2023) Analysis and modelling of laser-micro EDM-based hybrid micro milling on stainless steel (SUS304) using box Behnken design. In: IIUM engineering congress proceedings

Chapter 4
Conclusions and Future Trends

In summary, the exploration of laser-micro-EDM-based hybrid micromachining has uncovered a profound synergy between two distinct yet complementary machining techniques, offering a robust platform to meet the evolving demands of micromanufacturing. This study, through a systematic investigation into fundamental principles, process optimization strategies, and experimental validations, has furnished a comprehensive understanding of the capabilities and potential applications of this hybrid approach.

While the nanosecond LBMM process exhibits rapid machining capabilities, it suffers from compromised machining quality attributed to factors such as the large heat-affected zone (HAZ), recast layer, and tappered structures. The transition to ultrashort LBMM processes employing picosecond and femtosecond pulses presents a viable solution to these issues. However, the considerable initial investment required for ultrashort LBMM machines and their applicability primarily to low aspect ratio microstructures pose significant challenges. Conversely, the micro-EDM process boasts high-quality finished products with minimal HAZ and recast layer, as well as significantly reduced tappered structures compared to LBMMed microstructures. Nevertheless, its material removal rate (MRR) is orders of magnitude slower than that of the LBMM process. By harnessing the unique advantages of laser ablation and microelectro discharge machining (micro-EDM), laser-micro-EDM hybrid micromachining has transcended the limitations of conventional micromanufacturing techniques, empowering engineers and researchers to fabricate intricate microstructures with unprecedented precision, complexity, and surface integrity.

The experimental study detailed in this work demonstrates how the effectiveness of the laser-micro-EDM hybrid micromachining process can be finely tuned by adjusting LBMM process parameters [1]. Additionally, guidelines have been established to optimize laser parameters to minimize the final EDM process time. A methodology has also been outlined for positioning the EDM tool on pre-machined LBMMed structures utilizing an on-machine camera system. Furthermore, a dual-stage artificial neural network (ANN) model has been developed and extensively

T. Saleh et al., *Laser-MicroEDM Based Hybrid Micromachining*,
Manufacturing and Surface Engineering, https://doi.org/10.1007/978-981-97-8374-8_4

discussed in Chapter Three. In the first stage, laser parameters are inputted (laser power, scanning speed, loop count, and pulse repetition rate) to predict outputs such as entry and exit areas of machined holes, HAZ, and recast layer area. These predicted outputs are then utilized as inputs for the second-stage model to predict micro-EDM outputs such as machining time and tool wear, with the model demonstrating reasonable accuracy when evaluated with test datasets.

To fully harness the potential of LBMM-micro-EDM-based hybrid processes, integration of the two techniques into a unified setup is imperative. This work proposes a patented design for such a unique machine, complete with relevant on-machine location and measurement systems, facilitating seamless integration and optimization of the hybrid process.

In conclusion, the exploration of laser-micro-EDM-based hybrid microma-chining has unveiled a remarkable synergy between two distinct but complemen-tary machining processes, offering a robust platform for addressing the evolving demands of micromanufacturing. Through a systematic investigation into the funda-mental principles, process optimization strategies, and experimental validations, this study has provided a comprehensive understanding of the capabilities and potential applications of this hybrid approach.

4.1 Future Trends

Looking ahead, the future of laser-micro-EDM-based hybrid micromachining is poised for even greater advancements and innovations:

Material Diversity and Compatibility: Expanding the range of machinable mate-rials to include emerging advanced materials such as metal matrix composites, shape memory alloys, and 2D materials will enable the fabrication of microstructures with tailored properties and functionalities, unlocking new frontiers in material science and engineering.

3D Micromanufacturing: Pushing the boundaries of laser-micro-EDM hybrid micromachining into the realm of three-dimensional (3D) micromanufacturing will enable the fabrication of complex microstructures with hierarchical features, gradi-ents, and multi-scale architectures, revolutionizing fields such as microfluidics, photonics, and metamaterials.

Micro-Optics and Photonics: Leveraging the precision and flexibility of laser-micro-EDM hybrid micromachining for the fabrication of micro-optical components, photonic devices, and integrated photonics platforms will enable the development of compact, lightweight, and high-performance optical systems for applications ranging from telecommunications to biophotonics.

Microadditive Hybridization: Integrating additive manufacturing techniques such as micro-3D printing and laser-based direct writing with laser-micro-EDM hybrid machining offers a hybrid additive-subtractive approach for fabricating complex microstructures with enhanced design freedom, material versatility, and process efficiency.

Smart Manufacturing and Industry 4.0: Embracing the principles of smart manufacturing and Industry 4.0, laser-micro-EDM hybrid micromachining systems will evolve into interconnected, data-driven manufacturing ecosystems that enable real-time process optimization, predictive maintenance, and adaptive manufacturing strategies for enhanced productivity, quality, and sustainability.

Cross-Disciplinary Collaborations: Fostering interdisciplinary collaborations between researchers, engineers, and industry stakeholders from diverse fields such as materials science, optics, robotics, and artificial intelligence will catalyze innovation and drive the development of next-generation laser-micro-EDM hybrid micromachining technologies with transformative societal and economic impacts.

In summary, laser-micro-EDM-based hybrid micromachining represents a paradigm shift in micromanufacturing technology, offering a versatile, scalable, and future-proof solution for meeting the growing demand for miniaturization, complexity, and functionality in advanced microsystems and devices. By embracing a holistic approach that integrates cutting-edge technologies, advanced materials, and collaborative partnerships, the journey towards realizing the full potential of laser-micro-EDM hybrid micromachining promises to redefine the boundaries of micromanufacturing excellence in the decades to come.

Reference

1. Rashid MAN, Saleh T, Noor WI, Ali MSM (2021) Effect of laser parameters on sequential laser beam micromachining and micro electro-discharge machining. Int J Adv Manuf Technol. https://doi.org/10.1007/s00170-021-06908-8